Pesticides in Crop Production

Pesticides in Crop Production

Edited by:

Darshna Rathod

Delve
PUBLISHING

www.delvepublishing.com

Pesticides in Crop Production

Darshna Rathod

Delve Publishing
224 Shoreacres Road
Burlington, ON L7L 2H2
Canada
www.delvepublishing.com

Email: orders@arclereducation.com

ISBN: 978-1-77407-714-6 (Hardcover)

ABOUT THE EDITOR

Dr. Darshana Rathod, is presently working as Assistant Professor, Department of Agricultural Entomology, College of Agriculture, Parul University, India. She graduated from the Anand Agricultural University, with doctorate in Agricultural Entomology in the year 2017. She was Awardee of National Fellowship for pursuing Ph.D. sponsored by University Grants Commission. She was also awarded by Dr. R. C. Patel award for the best Research done in Ph.D. by Plant Protection Association Group of Gujarat. She has authored a number of national and international publications.

TABLE OF CONTENTS

LIST OF FIGURES

PREFACE

The pesticides can be called agrochemicals and they play an important role in food production. The population of the entire world depends on crop production to meet the food demands and Increased crop production demands quality pesticides that helps in healthy crop production. These chemicals can control disease vectors such as insects and microbes which can spread diseases. There are some advanced technologies such as biotechnology, crop genetics, formulations and precision agriculture which were proved to be effective in healthy crop production. There are some groups of chemicals such as strobilurin fungicides, neonicotinoid insecticides, imidazolinone herbicides and sulfonylureas which are considered crucial preparations in this century. There are some pesticides that are classified as safe. The production of safer pesticides can be a challenge to scientists working on various chemical formulations. Higher costs can be incurred whilst producing these pesticides. Most of the costs will be in the research and development with permissions and registrations. This cost barrier has become the biggest hurdle in pesticide research. Due to the high costs needed for the registration and re-registration of pesticides *i.e.* triazine herbicides, chloracetanilide, carbamate and organophosphate insecticides the production of pesticides was slowed down.

The latest pesticides were designed to protect the crops from pests. The action of pesticides on the crop plants must be kept at a minimum. The environmental concerns with these pesticides have become popular. Some of these pesticides can persist in the environment for long periods. Overall, the pesticide industries are challenged due to excessive regulations in their use and storage. It is important to find a balance between the pesticide actions and their environmental persistence. Most of mankind is based on crops and soils which are fertile. Some of these pesticides can be harmful to humans and they must be avoided. There are some chemical agents that can protect the crops from several pests and diseases andthey are called crop protection agents. These agents can enhance soil productivity and they can even eliminate weeds. Due to the evolutionary process, most of the microbes become resistant to these pesticides. These chemicals are being used all over the world on a massive scale. The pesticides have given humans several benefits. These benefits are accompanied by the food preservation, crop protection, and regulation of the insect vectors. The pesticides were classified into many classes based on their biological actions.

There are many regulations that can restrict the use of pesticides in several countries. Fungal diseases can damage most of the crops all over. Most of the crops can serve as a host for the fungal members. This fungal species can produce immense damage to the quality and quantity of the crops. In order to control these fungal infestations in the crops fungicides were developed. In the last decade, several crop communities have faced problems with fungal infections. Some of the severe problems were generated due to the excess growth of the weeds. The main problem with the weeds is that they compete with the actual crops and hence the crops of our interest will get damaged. Herbicides were developed in the industries to combat these weeds. Genetically engineered crops can help us to use herbicides. Nematicides are being used to kill the nematodes which attack the plant species.

This book aims at explaining different types of pesticides and their modes of action on plant pests such as insects, nematodes, and weeds. The resistance produced due to different mechanisms is discussed in this book. This book can help graduates and engineers who would like to pursue a career in the agriculture sector or relevant areas of research.

INTRODUCTION

The pesticides are the chemicals which that can be used to damage or kill the pests which produce cause enough damage to the crops. These chemicals can interfere with the physiological and biochemical processes. They often target on the pests but most of the pesticides are harmful to humans and wildlife(reference). Some of the pesticides can be lethal and we can classify the toxic levels according to their lethal nature. The pesticides can be acute and chronic in nature. Their persistence in the environment can be based on many factors. Some pesticides can be active in low temperatures whereas some work well in higher temperatures. There are several side effects of the pesticides. The pesticides were used since for ages and they may not be treated as novel chemicals. During the World Wars, chemicals used as pesticides were increased so that the crop protection was possible(reference). Farmers still are using Tar oil to regulate the growth of aphids even at the egg stage(reference). In 1932, Dinitro-orthocresol was used to regulate the weed growth. In the Second World War period, DDT was produced in heavy quantities in Switzerland whereas in Germany, organophosphorus compounds were produced in huge quantities(reference). In the 1960s, there was a drastic development of systemic fungicides. Some of the systemic fungicides are metalaxyl, benomyl, hexaconazole, triarimol and ethirimol. Metalaxyl is an acylalanine derivative and hexaconazole is classified as triazole. Triarimol and ethirimol comes under the pyrimidines and benomyland itis classified as benzimidazoles. The evolution of herbicide production was dominant during 1955 and 1960. In Britain, quartenary quaternary compounds were developed and in Switzerland, triazines were developed. Some of the important compounds produced are sulphonylureas like chlorsulfuron, diphenyl ethers such as acifluorfen and bromoxynil which is classified as hydroxybenzonitriles. Some of the foliar herbicides are sethoxydium and diclofop-methyl.

- The weeds can be regulated with the aid of these herbicides. Weeds are the plants which that can damage reduce the growth of the crop plants. Due to weeds, the economy of several countries will be dwindleddwindle. Weeds can interfere with theecosystems. They can and also spoil the water resources. Some of the weeds can be toxic to both animals and humans(reference). These plants can harbor diseases and a variety of pests. They can even damage crop harvesting machines.Fungicides can be any biological agents or chemicals which can diminish the fungal growth. These agents can also kill the fungal spores. Mostly the contemporary fungicides can inhibit the fungal growth. Fungi can damage crops and so there will be higher economic losses. The fungicides can be useful in agriculture and they can even control the fungal infections in animals(reference) . Fungicides can also regulate oomycetes. Fungicides can be classified as contact fungicides, translaminar fungicides and systemic fungicides.There are more than 10,000 insect species which are involved in damaging crops. Out of these, only 700 species create much damage to the crops. In World War II, some of the insecticides used in the combat were DDT and other synthetic organic insecticides. Organochlorines have chlorine, hydrogen and carbon groups. They are known as chlorinated hydrocarbons or chlorinated insecticides. Diphenyl apliphatics are considered as the oldest group of the organochlorines. Some of the compounds such as methoxychlor, chlorobenzilate, ethylanethylene, dicofol, DDD and DDT were classified as diphenyl aliphatics.
- The nematodes are invertebrates which are mostly non-segmented. Their body exhibits bilateral symmetry. There They will be havingpossess a body cavity and a well- developed digestive system. They don't have circulatory and respiratory systems. The body wall is made up of musculature, cords which are longitudinal and multiple cuticles. Close to the pharynx, there will be presence ofis a nerve ring. The excretory system will eliminate all thetargets wastes. Well The well- defined osmoregulatory system is seen in these nematodes. These systems are based on the species and the developmental stages they are in(reference). The pesticidal resistance is explained in the last chapters of this book.

Write more about of this book. What you are going to present here. Why this book is important and why people should read this?? Give references of books published earlier.

1

Introduction to the Pesticides

CONTENTS

1.1 INTRODUCTION

Chemicals were used to regulate the growth of insects since classical Greece. In Rome, farmers have used burningsulfur as fumigants (reference). Arsenic was used as a chemical that can kill insects. Olive oil and soda were also used to control the insect populations. Mostly they were used to protect legumes that were damaged due to insect attacks. Arsenicals were used as pesticides in China. At the start of the nineteenth century, many farmers started using soap and pyrethrum to control the insects. In the year 1896, grape farmers used the Bordeaux mixture. When this chemical was used on the dicotyledonous weeds and cereals, there was a lesserdamaging effect on the plant but the weed population got reduced. Organomecury was used to dress the seeds so that they were contamination free. The uses of the chemicals made of organomercury were evident in the First World War period. During these World Wars, the chemicals used as pesticides were increased so that the crop protection was possible. Farmers still look at Tar oil to regulate the growth of aphids even at the egg stage. In 1932, Dinitro-orthocresol was used to regulate weed growth. In the Second World War period, DDT was produced in heavy quantities in Switzerland whereas, in Germany, organophosphorus compounds were produced in huge quantities. In the 1960s, there was a drastic development of systemic fungicides. Some of the systemic fungicides are metalaxyl, benomyl, hexaconazole, triarimol and ethirimol. Metalaxyl is an acylalanine derivative and hexaconazole is classified as triazole. Triarimol and ethirimol come under the pyrimidines and benomyl is classified as benzimidazoles. The evolution of herbicide production was dominant during 1955 and 1960. In Britain, quarternary compounds were developed and in Switzerland, triazines were developed. Some of the important compounds produced are sulphonylureas like chlorsulfuron, diphenyl ethers such as acifluorfen and bromoxynil which is classified as hydroxybenzonitriles. Some of the foliar herbicides are sethoxydim and diclofop-methyl.

Several farmers got benefitted in Europe and North America due to the development of pesticides that were not toxic to the crops but regulate weed populations. Most of the diseases and pests choose the monocultures and it was evident in intensive agriculture (reference). Pesticides have also brought resistance in the weed plants which resulted in huge losses. There were several ecological problems due to pesticide use. 2/3rd of the herbicide market is in the USA. The agricultural productivity was increased with the combined use of irrigation, fertilizers, mechanization and pesticides. The chemicals can induce two effects:acute effects and chronic effects. Acute effects generally occur immediately after the exposure of the weeds to the poisons or chemicals. The acute effect prevails in response to a single dose of a chemical. This effect is based on the method of application, toxicity and the organisms on which a specific chemical is applied. Sodium chloride will produce less effect on animals than arsenious oxides. We can quantify the acute effects as they can yield noticeable signs and symptoms. Most of the effect is biological or physiological in nature. If a chemical is applied repeatedly in minute doses which are non-lethal, we can expect a chronic effect. Some of the consequences of chronic effects are necrosis, brain damage, lung cancer and silicosis. We can observe necrosis in the kidneys and liver. We can measure the acute toxicity in terms of lethal dose which can be designated as LD_{50}. It means that about 50% of the organisms under study will die when a particular chemical is applied. LC_{50} is another crucial concept to explain the chemicals present in air or water which can bring 50% mortality of the species existing in that area. We can use ED_{50} and EC_{50} to explain the effective dose and effective concentrations. We can use I_{50} which explains 50% of inhibition of the species. Edson et al. in 1966 and Fairchild in 1977 provided acute toxicity data of several pesticides. If the LD_{50} values are less, we can say that the chemical is highly toxic and if LD_{50} value is higher, then the chemical is said to be less toxic. The toxicity of any chemical can be expressed in parts per million or ppm. It is expressed as a function of body weight. Du Bois and Geiling in 1959 provided a scale for assessing the toxicity levels. According to this scale, if $LD_{50} < 1$mg/kg, we can say that the substance used is extremely toxic. In case the LD_{50} values range between 1 to 50 mg/kg, then we can call such a substance as highly toxic. If this value is between 50 to 500 mg/kg, we can call it moderately toxic and if the value lies between 50 and 500 mg/kg, we can express it as slightly toxic in nature. Acute oral LD_{50} values are represented in figure1.1.

Insecticide	No. Species Tested	HD_5 (mg/kg Body Weight)	LD_{50}(mg/kg Body Weight)
Aldrin	12	1.15	19.8
Chlordane	4	0.09	62.3
Chlordecone	2	26.4	220
DDT	5	123	1330
Dicofol	3	72.4	681
Dieldrin	16	4.15	35.1
Endosulfan	6	9.53	52.4
Endrin	11	0.75	1.78
HCH (technical mixture)	1	12.5	N/A
Heptachlor	7	3.47	125
Lindane	11	10.5	90.8
Methoxychlor	4	292	N/A
Toxaphene	11	10.4	70.7

N/A—Not available.

∗The table gives the number of bird species tested as well as the median LD50 value and the HD5. This is the value calculated (with 50% confidence)

Figure 1.1. Acute oral LD_{50} values.

Source:https://www.researchgate.net/publication/228041117_Avian_Species/figures

Mevinphos has an LD_{50} value of 3 mg/kg and it is an insecticide. Fungicide named Ethylmercury chloride has an LD_{50} value of 30 mg/kg whereas herbicide Dinoseb has an LD_{50} value of 50 mg/kg. McCann and Ames in 1976 devised a mutagenicity test to explain the toxicity of pesticides on salmonella. This test was used by Wildemauwe et al. in 1983 to describe the activity of the pesticides. He explained the mutagenicity of toxaphene, pirimiphos-methyl and dicofol.

1.2 CLASSIFICATION

The World Health Organization or WHO has classified acute pesticide toxicity. WHO has classified the acute pesticide toxicity into four main classes. They are Ia, Ib, II and III. Class Ia can be calledan extreme hazard, class Ib is a high hazard, class II is a moderate hazard and class III is a slight hazard in nature. This classification is based on the LD_{50} values obtained when experiments were done on rats to get LD_{50} values. The rats were subjected to the pesticides orally and percutaneously. Environmental Protection Agency or EPA has classified pesticide acute toxicity. EPA classified the acute toxicity into classes I, II, III and IV. The pesticides are classified as biological and chemical subgroups. Federal Insecticide, Fungicide and Rodenticide Act or FIFRA defines pesticides as a substance or mixture which can mitigate any pest or kill any pest. The pesticides can be natural products such as derris root, tobacco, rotenone and nicotine. The copper fungicides can be used as protecting agents that can regulate fungi. There are several pathways where humans can get affected due to pesticides. The exposure can be oral, percutaneous and inhalation routes. In the case of veterinary or human medicine, occupational exposure, vector control, pesticide for public hygiene, horticulture or domestic applications can influence humans via oral, percutaneous and inhalation routes. The water and food residues can influence humans through the oral route. The classification of the pesticides is given in figure 1.2.

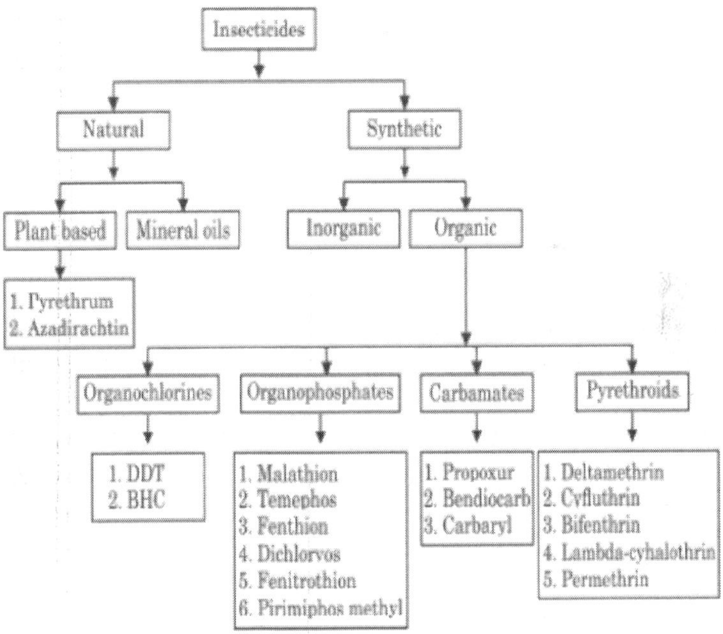

Figure 1.2. Classification of pesticides.

Source:https://www.ijcmas.com/8-3-2019/Rajveer%20Kaur,%20et%20al.pdf

1.3 INSECTICIDES

The insecticides are made up of organochlorines and their use in agriculture is crucial. In addition to this they can control malaria. The chlorinated hydrocarbons are classified into cyclodiene, lindane, toxaphene, DDT and chlordecone/ mirex. These chemicals will affect neuronal activities in humans and other animals. The isobenzan, dieldrin and endrin are considered toxic than DDT. We can say that they are ten times more toxic than DDT. These insecticides can affect humans through the skin. Humans can get affected via oral and respiratory routes. They can get absorbed by the intestines as the food materials will comprise of fat and fiber components. These insecticides were discouraged due to harmful properties. These chemicals can deposit in the fat tissues and their concentration is higher in these tissues than other organs. DDT and DDE can be stored in the fat tissues and methoxychlor can be seen in low concentrations in the fat tissues. This is the reason for banning DDT and DDE. They can affect the electrophysiological properties of the nervous system. There will be an imbalance in the sodium and potassium ions. This will upset the calcium transport. The organochlorine insecticides and DDT can affect the liver in mice and other animal models. The liver damage is linked with enzymatic and microsomal activities. DDT can enlarge smooth endoplasmic reticulum. Due to altered mitochondrial activity, there will be an increase in the number of lysosomes. Because of these effects, the Golgi bodies will get damaged. Most of these insecticides can disrupt the endocrine activities in the animal systems and the effect on humans is yet to be established.

HCH or Lindane can exist in eight steric isomeric forms. Lindane can be used in the form of powders, sprays, aerosols and concentrates. These chemicals can regulate the growth of soil pests, wireworms, rice insects, cotton insects and grasshoppers (reference). Lindane is being used in the seed protection processes and poultry. The gamma form of hexachlorocyclohexane is called as lindane. We can use lindane as pediculocide and scabicide. The HCH can be produced through various routes and the transformation of lindane in humans can take several pathways. Lindane can be converted into chlorobenzene through dechlorination. Eventually, the chlorobenzene can get converted to benzene. HCH compounds can cause epileptic effects in humans. The organochlorine pesticides can produce chronic oxidative stress. It was reported that lindane can damage seminiferous tubules. Vaporizer pellets can comprise of these compounds and this can be toxic to children. The classification of insecticides is represented in figure 1.3.

Insecticide	Chemical compound	Commercial product
Organophosphorus insecticides	TEEP - tetraethyl pyrophosphate	FYFANON 440EW
	Dianizion	ANTHON
	Parathion	BOVINOX
	Paraoxon	PROXOL
	Trichlorfon	DANEX
	Malathion	
Polychlorinated insecticides	Bischlorophenyl:	AZOTOX
	DDT	DITOX
	Metoxychlorine	TRIOX
	Derivatives cyklodiene:	ANOTEX
	Aldrin	CESAREX
	Dieldrin	CHLOROPHENOTHANE
	Ceptachlorine	DEDELO
	Endosulfan	DINOCIDE
	Chlordon	DIDIMAC
	Derivatives cycloparaffinic:	DIGMAR
	Hexachlorocyklohexane	GENITOX
	Lindane	GUESAPON
	Chlorinated terpenes:	IXODEX
	Camphenes	NEOCID
	Pinenes	R50
Carbamate insecticides	Carbanil	PRIMOR 500W
	Carbofuran	MARSHAL 250DS
	Primor	PRIFLOR AE
	Izolan	PRIMIX AL
	Aldikarb	PRIMIX AE
	Bindokarb	
	Thiamethoxan	
Neonicotinoids	Imidaclopirid	ACTARA 25WG
	Firponil	MOSPILAN 20SP
	Clothianidin	
	Natural:	
Pyrethroids	Pyrethrins	CYFLOK 50EW
	Jasmolin	K-OTHRINE
	Cynarin	BROS INSECT SPRAY
	Synthetic:	ABC
	Allethrin	AC
	Alphamethrin	INSEKTUM A 01 AL
	Bioresmethrin	SPRUZIT 04EC
	Cyfluthrin	DECIS 2.5EC
	Cypermethrin	
	Deltamethrin	
	Permethrin	
	Rosmethrin	
	Trousfluthrin	
	Fenvalerate	
Others	Calcium oxide	
	Boric acid	
	Trioxane	
	Nicotine	
	Zyklon B	

Figure 1.3. Classification of insecticides

Source:https://www.researchgate.net/publication/306191667_The_use_of_insecticides_to_control_insect_pests/figures?lo=1

Cyclodiens can exhibit neurotoxic effects on humans. Dieldrin can create less damage to humans than endrin. Alpha endosulfan, endrin, dieldrin, aldrin, heptachlor and chlordane can be used in agriculture (reference). Aldrin can produce a distinctive flavor when crops treated with this chemical are ingested. Dieldrin can be used in houses to regulate vectors. These insecticides can be absorbed in the intestines if dissolved in corn oil or peanut oil. Some of the cyclodiene chlorinated insecticides are chlordane, heptachlor, aldrin, dieldrin, endrin and endosulfan. Sulfate derivate can be considered as a metabolite of endosulfan. Dieldrin can be considered as a metabolite for aldrin. The neurotoxic activity of this chemical on the nervous system starts from the synapse. GABA receptor can be an important binding site for dieldrin. The picrotoxin binding site is an important area on which the dieldrin molecules can attach.

Toxaphene is an important insecticide that can be produced due to camphene chlorination. The chlorination of camphene must be in excess. It is considered powerful as this chemical can be excreted by the animals and mammals easily. Though it is considered powerful, the rate of acute toxicity is the same as lindane. This compound can be mutagenic. When ingested by humans we can observe symptoms like convulsions, jerking, nausea and mental confusion. The chromosomal abnormalities in the lymphocytes are predominant when ingested.

DDT is another pesticidethat was popular for its useful effects and harmful effects. This chemical was produced by Zeidler in the year 1992(reference). Methoxychlor, dicofol, chlorobenzilate and ethylanare considered as analogs of DDT. The absorption of DDT on the skin is much less. The powder comprising DDT can be observed in the upper portion of the respiratory tract. The spread of DDT in our body can be dominant in the lymph system. Eventually, DDT takes its path to the chylomicrons and then to the plasma proteins. The DDT concentrations can be observed in almost all the part of the body where fats are present. DDT can be excreted through bile. The nervous system can be the most affected. The closing of gates in the ion channels of the axiom complex can be affected by DDT. There will be a lowering of magnesium, potassium and sodium ions. The calcium and ATPase regulation at the surface of the axon can be disturbed due to DDT concentrations. This chemical keeps the ion gate ofNa channel open for long periods.

Anticholinesterase insecticides are considered as harmful as they attack the nervous system in both animals and humans. These insecticides can inhibit acetylcholinesterase or AchE. They act on the central nervous systems and peripheral nervous systems. AchE was isolated from several animal species. It was found that these molecules belong to the serine hydrolases family. The 3D structure of AChE was studied with the aid of X-ray crystallography and NMR (reference). Based on these investigations, we can say that this enzyme has three important regions that are functional. There isan anionic site on the surface. There is acavity that can link the active site with the surface area. The catalytic triad is present and this will help in the mechanics of the site. AchE is classified into two classes. They are heteromeric subunits and homomeric oligomers. The main work of AChE is to terminate the impulse which can happen if Ach is released. In the process of AChE hydrolysis,Ach can be divided into acetate and choline. In normal conditions, there will be 10000 Ach conversions.

There are two important classes of insecticides that can influence AChE. They are organophosphorus compounds or OP compounds and carbamates. These compounds can inactivate AChE. The compounds can alter the activity of AChE. They can attach to AChE covalently especially at the serine amino acid regions. The carbamate compounds can be considered as anti-AChE insecticides. These insecticides can control a wide variety of pests. These compounds can get degraded easily and the preparation of them is quite easier. Urethane can be considered as a carbamate functional group. We can consider these compounds as ester amides. These groups must be derived from the oximes, naphthols, phenols and alcohols. The carbamates can take two pathways when reacted with water. In case of higher pH, the carbamates can react via ester bond. In low pH conditions, the cleavage will start from the amide bond. The carbonate monoesters and N-carboxylic acids can be unstable. The carbamates can undergo hydrolysis resulting in carbon dioxide, amine and alcohol. The hydrolysis process can occur both in acidic and alkaline conditions. The ester position can be used to form oximate and aryloxy groups.

The organophosphate compounds are also called OP compounds. We can include nerve agents and insecticides as OP compounds. These compounds can constitute organic acid halides and esters. The OP compounds are dominated by organophosphinates and other compounds. The parathion, malathion, diazinon, chlorpyrifos and acephate can be considered as important OP insecticides. Most of these OP insecticides cannot be chiral in nature. The compounds such as acephate and malathion can have asymmetric centres. These compounds exist as enantiomers and stereoisomers. These OP insecticides can exhibit P=S bonds. This bond is known as thionate bondand these compounds can be called phosphorothionates which are a subgroup of OP compounds. These compounds can be in an inactive state mostly and they act as anti-AChE agents. These reactions of OP compounds can be catalyzedrapidly by cytochromes p450. Because of these catalytic reactions, there will be the formation of P=O bonds. This bond is highly polarised. There are other compounds that can be more reactive than anti-AChE agents and they are namedoxons. The oxons can get isomerized to form phosphorothiolates. There will be the formation of thiolester and oxon in the same molecule. There will be nucleophilic displacement due to these reactions.

Due to thermal and chemical treatments, there will be thiono-thiolo rearrangements. The isomerizations generally occur at 180°C. These reactions can also occur when reacted with alkyl halides. In some cases, these isomerizations can occur in room temperatures.

The oxons can alter the AChE activity and can detoxify the carboxylesterases. With these reactions, we can say that it has an anti-AChE effect. The malathion compound can have chiral centres. This centeris located at the ligand carbon. There will be the formation of an extra chiral centerespecially at the P and this is formed due to the isomerization process. There will be isomerization between 4 stereoisomers. The inhibition of the AChE compound by the carbamates can be explained with the catalytic hydrolysis process.

1.4 PYRETHRINS

The heads of the chrysanthemum flower have certain insecticidal properties and this property was discovered 2000 years ago. Ruzicka and Staudinger in 1924 explained the chemical basis of this property. These insecticidal activities can be attributed to the pyrethric acids, chrysanthemic acids and their association with jasmolone alcohols, cinerolone and pyrethrolone compounds. In general, these pyrethrins cannot be stable but we can use these compounds as insecticides. Bioremethrin is considered as the first compound of this kind. Due to the addition of a cyano group on the alcohol group, there will be the formation of 8 isomers. These pyrethroids can be sold as racemic mixtures which can be named as permethrin and some exist as single isomers and they are called deltamethrin. There was the production of synthetic pyrethroids in large scales and they were used as agricultural pesticides. These synthetic chemicals are resistant to ultraviolet radiation and they are stable and can be applied to the crops without much problem. The pyrethroids and pyrethrins can be used as domestic insecticides and the problems associated with human health are at a minimum. We can use these compounds to regulate disease vectors which are insects. Pyrethrins and pyrethroids can exhibit low volatility rates. When we inhale these vapors, the aerosol droplets can be inhaled which are harmful to humans and other animals. Due to the oxidative stress generated on the application of these compounds, these pests die (reference). The oxidation stress will be due to decreased concentrations of glutathione. In the case of chronic exposure, humans can feel discomfort and irritations may persist. In mammals, the pyrethroids can target chromosomes, gap junctions, mitochondria (ATPase), lymphocytes, nicotinic receptors, GABA channels, voltage channels, membranes, noradrenaline and protein phosphorylations. The toxic levels can be increased from 10^{-13} to 10^{-4}M. These compounds can be called functional toxins. They can cause hyper-excitation in mammals and animals. Carcinogenicity and genotoxicity can be observed due to their exposure. The nervous system can get affected much. The classification of pyrethrins is shown in figure 1.4.

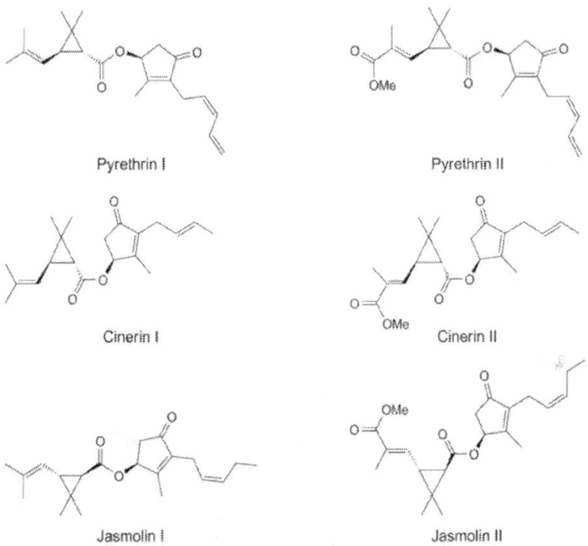

Figure 1.4. Classification of Pyrethrins.

Source:https://www.researchgate.net/publication/236049582_Chemical_diversity_of_the_natural_populations_of_Dalmatian_pyrethrum_Tanacetum_cinerariifolium_TREVIR_SCHBIP_in_Croatia/figures?lo=1

The OP compounds can inhibit AChE. The serine group present in the active site can be disrupted due to covalent modification. The inhibitory pathways can be with various processes. There are two major types of reactivation. They are mainly oxime induced and spontaneous reactivation (reference). When phosphoserine bond is cleaved with water, we can expect a spontaneous reactivation. If oxime antidote is utilized to break the phosphylserine bonds, then we can call it as oxime induced reactivation. Another mode of reaction is the non-reactivation type. If all the activities linked with the phosphorylated AChE is failed to restore, we can call such type of reaction as a non-reactivation type. We can consider aging as an important mechanism involved in the non-reactivation type. In this aging mode, there will be alkylation of phosphoryl groups.The oximes can exhibit anti-AChE activity and the spontaneous reactivation can result in carbamoylatedAChE. The pyetherins and pyrethroids can be classified into types I and II. Type I includes tetramethrin, pyrethrins, permethrin, kadethrin, bioresmethrin, bioallethrin and bifenthrin. Type II includes fenvalerate, deltamethrin, cypermethrin, cyhalothrin and cyfluthrin.

Type I and II pyrethroids can be employed to activate adrenalin in rats. The pyrethroids which belong to the type I can cause poisoning syndrome. There will be the production of Na tails. These tails have shorter time constants. The toxic nature of the pyrethroids can be compared with the toxicity of DDT. This was explained by Aldridge and Verschoyle in 1980. Type II pyrethroids can produce complex poisoning syndrome. These chemicals can affect diverse tissues. One of the important effects of these compounds is acute excitation. These compounds even in slight doses can result in axonal degeneration. Neurotoxicity is a common effect seen in humans. Deltamethrin and permethrin can produce behavior changes in adults.

1.5 MISCELLANEOUS INSECTICIDES

The miscellaneous insecticides can be extracted from the plants and other organisms. The active substances present in these insecticides are methoprene, tebufenozide, diflubenzuron, indoxacarb, fipronil, imidacloprid, spinosad and abamectin. Neuroactive insecticides can be called as miscellaneous insecticides. These compounds have antibiotic formulations which belong to oxadiazine and phenylpyrazole families. Some of the miscellaneous insecticides are abamectins and ivermectins. Both of them come under avermectins. These compounds can be used to control insects, mites and fungi. These compounds were formulated in 1970 from *Streptomyces avermitilis*. These compounds will influence neuromuscular transmission via GABAergic neuronal systems. There will be paralysis in both nematodes and arthropods due to the effect of these insecticides.

Abamectin can be used for agricultural use. The avermectins can get absorbed through the skin tissues. In rare cases, these compounds can take an oral route. It is obvious that these avermectins doses not possess mutagenic or genotoxic potentials. Spinosad is produced as a result of fermentation. This compound is produced due to *Saccharopolyspora spinosa*. This species belongs to bacteria which are classified as actinomycete bacterium. It has immense insecticidal activity. This compound comprises of two sugars, an amine group and a hydrophobic ring which is complex in nature. This compound consists of many spinosyns. The spinosyns isclassified into A and D and they exist in the ratio of 7:1. The neurotoxicity is seen in the insects upon exposure to spinosad compounds. We can observe coordination loss and tremors along with other effects. This compound exhibits less toxicity if administered dermally or orally. The oral administration can be via inhalation. This compound can produce irritation in the eyes. Spinosad did not exhibit mutagenic or genotoxic effects in lab conditions. In humans, there was no poisoning or any other noticeable reactions. Chloronicotinyl insecticides can be important miscellaneous insecticides. Imidacloprid is the best example of this type of insecticides. The insecticidal properties can be attributed to the presence of nicotine. Nicotine is mostly extracted from the *Nicotiana tabacum* plant. This compound exhibit high levels of toxicity in mammals. Due to this activity, nicotine use as an insecticide was stopped. Nicotine was replaced by a neonicotinoid insecticide called imidacloprid in 1992. This compound can regulate insects, especially those which are sucking type (reference). In the case of cockroaches, there will be leg tremors, leg strength weakening and death.

Imidacloprid can act rapidly and its absorption rate in rats is higher. The gastrointestinal tract of rats can be damaged due to the activity of this compound. This compound exhibits acute toxicity in mammals. There are no signs of dermal irritation or inhalation toxicity. Phenylpyrazole is another important insecticide. It is called fipronil. This compound belongs to family phenylpyrazole. It is being used for both agricultural and non-agricultural uses. This insecticide can act on the GABA chloride channels. There will be damage to the central nervous system. This compound did not show any mutagenic or genotoxic effects in lab conditions. In humans, there were no harmful effects found. Oxadiazine insecticides can be called as indoxacard. It is an active insecticide. It can exist as both inactive and active isomer of insecticide. These isomers are exist in the ration of 3:1. In some cases, the ratio can be 1:1. These compounds can block sodium channels. These compounds are effective against lepidopteran insects. The arthropods and bees can be controlled with this compound. This compound can be metabolized easily. There will be the formation of aniline metabolites due to the excess metabolism of this chemical. In rats, there was 10% of absorption in lab conditions. It is not carcinogenic in nature. In animals, we can observe decreased food consumption and hence lowered body weight. There was negligible genotoxicity in rats. In humans the toxicity levels incurred due to this chemical is negligible.

Insect growth regulators are important in controlling the growth of insect pests as they intercept all the growth stages of insects. Some of them are juvenile hormone analogs, ecdysone agonists, chitin synthesis inhibitors. Diflubenzuron is one of the chitin synthesis inhibitors. It functions as both insecticide and acaricide. This compound can control the larva stages. We can use this insecticide on ornamental plants, forests and agricultural fields to regulate insects. We can use this compound on cattle, ornaments, cotton and others to regulate the populations of gypsy moth and mosquito larvae (reference). This type of insecticide can function as a chitin synthesis inhibitor. These chemicals can regulate the growth of flies, weevils, rust mites and leaf- eating larvae. It also exhibits ovicidal action. The chitin concentration of the cuticle will get diminished due to the activity of this insecticide. Teflubenzuron can be compared with this chemical as both of these chemicals can the same type of action. Diflubenzuron may not be neurotoxic as none of the observations were observed in lab conditions. Genotoxicity was not seen with this chemical.

Tebufenozide belongs to the class of Ecdysone agonists. This chemical is non -steroidal. This chemical belongs to insect growth regulators. It shares the same characteristics as 20-hydroxyecdysone. This chemical is a molting hormone seen in invertebrates. This chemical can alter behavioral effects in most of the insects. It is used in inducing larval molt which is a natural process. Tebufenozide will mimic this chemical and result in premature larval growth so that the insects will die. The ecdysis process will be failed completely in the mosquitoes. There are 4four aquatic arthropods that will be affected by this chemical.

1.6 FUNGICIDES

The fungicides can be used to prevent fungal growth in agriculture and horticulture crops. These fungicides can also eradicate fungi in the inpowderries. The fungicides have both organic and inorganic substances. We can classify fungicides based on their modes of application. They are primarily classified into soil fungicides, foliar fungicides and dressing fungicides. According to their mode of action, the fungicides are classified into protective fungicides, post-infection fungicides and post- symptomatic fungicides. The protective fungicides will have foliar effects and sporicidal effects. The post-infection fungicides can be used to kill the mycelium. The post -symptomatic eradication fungicides can kill spores and mycelium. Phillips in 2001 has proposed certain properties to explain fungicides. The fungicides may limit the biodegradation on the plant surface. They may exhibit higher penetration rates for mycelia and spores. They may exhibit minimum phytotoxicity and low ecotoxicity. Azoles are a group of fungicides which include triadimenol, triadimefon, terbuconazole, penconazole, hexaconazole, fenbuconazole, etridazole, diniconazole and cyproconazole. Chloroalkylthiocarboximides can be effective fungicides. They include folpet, captafol and captan. Benzimidazoles or thiabendazoles comprise of fuberidazole, imazalil, thiophanate-methyl, thiabendazole and benomyl compounds. Carbamic acid derivatives can be classified into ethylene bisdithiocarbamates and dithiocarbamates. The ethylene bisdithiocarbamates includes zineb, mancozeb and maneb. The dithiocarbamates include ziram, thiram, ferbam and metam-sodium. The halogenated substituted monocyclic aromatics include dichlorophen, dinocap, pentachlorophenol, quntozene, hexachlorobenzene, dichloran, chloroneb, tecnazine and chlorothalonil. Some of other fungicides include morpholines, oxathiins or carboxanilides, organophosphates, piperazines, metallic compounds and other miscellaneous compounds. The classification of fungicides is represented in figure 1.5.

Chemical Name	Trade Name	Toxicity Classification	96 hr LC_{50} (mg/l): Rainbow Trout
Benomyl	Benlate	High	0.2
Captan	Agrox, Captan, Captec	Extreme-moderate	0.06
Carboxin	Vitavax	High-moderate	>0.1
Chlorothalonil	Bravo, Daconil,Terlanil	High	0.3
Coppersulfate	Basicop,Bluestone	High-moderate	0.14
Fenarimol	Rubigan	High	0.2
Fosetyl -Al	Aliette	Minimal	428
Iprodione	Rovral	Moderate	4
Mancozeb	Dithane, Fore, Manzate	High-moderate	2 (48 hr)
Maneb	Maneb, Manex	High-moderate	2 (48 hr)
Metalaxyl	Ridomil	Minimal	>100
Propiconazole	Alamo, Orbit,Banner, Tilt	High-moderate	0.9
Thiram	Thiram, Spotrete	High-moderate	0.1
Ziram	Ziram	Moderate	5 (5 hr):Goldfish

Figure 1.5. Classification of Fungicides.

Source:https://www.researchgate.net/publication/327845862_Ecotoxicology_of_Aquatic_System_A_Review_on_Fungicide_Induced_Toxicity_in_Fishes/figures?lo=1

The fungicides present minimum toxicity to humans and in some cases, these chemicals can result in carcinogenicity and genotoxicity. The halogenated monocyclic aromatics can have aromatic rings such as phenates and phenols. Some of the chemicals which are in this group are sodium pentachlorophenate, pentachlorophenol, quintozene, hexachlorobenzene, dichloran, chloroneb, tecnazineand chlorothalonil. These compounds can be metabolized and they can disrupt the oxidative phosphorylation process. Due to these reactions, animals and humans can have liver damage and hyperpyrexia. Chlorothalonil can be considered as a foliar fungicide. This fungicide is non-systematic in nature. It functions as a skin irritant and eye irritant. Tecnazene can work as a curative fungicide. It can also serve as a protective fungicide. We can use these fungicides in the form of dispersible powder, fumigants and granules. It has no capacity to irritate the skin in lab conditions. Chloroneb can function as a fungicide that can be systemic and applied as a broad -spectrum fungicide. This fungicide can be absorbed through roots. This can serve as non-irritant. Dicloran can be used as a protective fungicide. This fungicide can interfere with RNA transfer. We can control sclerotium, sclerotinia, rhizopus, monilinia and botrytis. It can disrupt the oxidative phosphorylation process. Hexachlorobenzene can be used against fungal spores. We can use this fungicide in seed treatment. We can use this fungicide in the form of dispersible powder or dry powder. This chemical does not exhibit any sort of genotoxicity. Quintozene can be used as a fungicide that can be in contact with soil and seed. We can apply this chemical to field crops and vegetables. We can use this fungicide as a wettable powder, suspension, granular formulation or in the form of dispersible powder. This chemical does not irritate the skin. There are some cases of dermatitis when this chemical is ingested or contacted via skin surface. We can observe a definite metabolic activation that is supported with the Ames test. When sprayed on plants, this chemical can be converted into pentachloroaniline and methylthiopentachlorobenzene.

Pentachlorophenol can be used to preserve the wood which can be infected with a fungal rot. We can observe some impurities such as polychlorinated dibenzofurans and polychlorinated phenols. We can use this chemical in the form of wettable powder and granules. We can use this fungicide in the form of oil -based liquids. It can be absorbed through percutaneous, oral and inhalation modes. We can observe dominant concentrations in both the kidney and liver. Pentachlorophenol can reduce the ATPase activity of mitochondria. In blood, milk and fat tissues we can observe pentachlorophenol. Dinocap can be used as a contact fungicide that can be used in both protective and curative functions. We can control the infection due to powdery mildew in vegetables and fruits. This fungicide can be used as a wettable powder and dispersible powder. This fungicide can disrupt the oxidative phosphorylation process. Occasionally there will be skin irritation in humans and animals. Dichlorophen can be considered as a contact fungicide. This fungicide can be used to regulate the moss and Fusarium species. We can use this fungicide in the form of soluble concentrate and emulsifiable concentrate. This fungicide can alter the oxidative phosphorylation reactions. Dithiocarbamate fungicides can be considered as esters of carbamic acid. They are considered as carbamate analogs. Compounds such as ziram, thiram, ferbam and metam can be included under dithiocarbamates. They can inhibit the activity of alcohol dehydrogenase which can be toxic. They can act as cholinesterase inhibitors. Metam sodium can decompose methyl isocyanate. This can function as a soil fumigant. This chemical can corrode skin. Ferbam can act as a foliar fungicide. It can work effectively on a blue mold on tobacco. We can use ferbam in the form of wettable granules. Thiram can be used as a contact fungicide. It can provide protection. It can be used to regulate botrytis on vegetables. We can control this species via seed treatment. We can use this fungicide in the form of suspension concentrates, flowable concentrates and dispersible granules or powders. This fungicide did not exhibit any carcinogenic effects (reference). Ziram can be used as a fungicide which can be classified as contact and foliar fungicide. We can use this fungicide in the form of liquids, powder and wettable granules. We can control the growth of fungi on vegetables and fruits.

Ethylene bisdithiocarbamates can be used as fungicides. The compounds such as zineb, trimanzone, nadam, mancozeb and maneb come under these fungicides. These compounds can create allergic reactions which can lead to dermatitis. The ethylene thiourea or ETU can work as a decomposition product or contaminant product. Maneb can be useful to control the fungal growth on the fruits (reference). This fungicide can be sprayed on the leaves and even we can use it in the seed treatment. We can use it as a wettable powder, suspension concentrate and as granules. We can consider Maneb as a skin irritant. In some cases, it can irritate eyes. Mancozeb is used to regulate several fungal diseases. Some of the fungal diseases which can be controlled by mancozeb are scab, rust downy mildew, leaf spot and blight. This fungicide can be used to treat seeds and on the leaves. We can use this fungicide as wettable granules and seed treatment. This chemical can irritate skin and eyes. Zineb can act on the leaf surface and regulate fungal populations. We can call this fungicide as a foliar fungicide. It can control fungal growth on vegetables and fruits. We can use it as wettable powders and dispersible powders. This can irritate the skin and eyes. Benomyl is a systemic fungicide. It can be absorbed by the roots and leaves. The process of absorption can be described as acropetal translocation. Some of the fungal members which can be controlled by this fungicide are basidiomycetes, fungi imperfecti and ascomycetes.

Thiabendazole can be classified as systemic fungicide. It has both curative and protective actions. This chemical can alter the movements of tubulin and so the process of mitosis can be inhibited. The fungal growth and development can be inhibited due to the disruption of mitosis. We can use this chemical as a post-harvest fungicide.

We can use this chemical via the foliar application and in the seed treatment. We can use this chemical in the form of wettable powder and soluble concentrate. Thiophanate methyl can be used as a systemic fungicide. It has both curative and protective actions. It can be absorbed via roots and leaves. This chemical can irritate the eye and skin. Carbendazim can inhibit the synthesis of beta- tubulin.

We can call it a systemic fungicide. It has both protective and curative functions. It can be absorbed via roots via acropetal translocation. It can control the growth of fungi in sugar beet and cereals. We can use this chemical in the form of a solution concentrate and as oil dispersible powder. It is safe and there will not be any irritation to the eyes and skin. Imazalil can be considered as an inhibitor in the biosynthesis of ergosterol. We can consider this chemical as a systemic fungicide.

It has both curative and protective actions. We can regulate several fungal diseases that infect vegetables and fruits. This chemical can be used in the form of soluble liquid and as emulsifiable concentrates. Fuberidazole can be called as a systemic fungicide. It can kill the Fusarium species. This chemical can inhibit mitosis. We can use this chemical in the form of dispersible powder and for the treatment of seeds (reference). It cannot irritate eyes and skin.

The chloroalkylthiodicarboximides can be called as R-SCCI compound. We can classify them as N-heterocyclics. In this group, we can observe folpet, captafol and captan. These chemicals are not toxic to humans. Captan can be considered as curative fungicide and protective fungicide. It will regulate the fungal population on the fruits. We can use this chemical to treat seeds. Some of the species which can be controlled are rhizoctonia, phoma and pythium. We can use this fungicide as a post- harvest chemical. We can use this fungicide in the form of wettable powder and as dispersible granules. This chemical can irritate eyes and skin.

Captafol can be considered as non-systemic fungicide. It has both curative ad protective actions. It can damage spores and we can consider it as a broad- spectrum fungicide. It can be used as seed protectant. It can irritate the eyes and skin.

Folpet can be considered as a foliar fungicide. It has protective action. We can use this chemical to control scab, leaf spot disease, powdery mildew and downy mildews. Rhizoctonia, pythium, alternaria, botrytis and gloeosporium species can be controlled with this fungicide. We can use this fungicide in the form of powder, granules and as dispersible powders. This chemical can act as a moderate irritant.

Azoles are considered as the most popular fungicides. The fungicides which are mostly used in regulating fungal diseases can be classified as triazoles. These are 5 ringed structures. Most of these fungicides can act as sterol demethylase inhibitors.

Because of this property, the synthesis of the cell wall is disrupted and hence fungal members will die. Cyproconazole can be considered as a sterol demethylase. We can classify this chemical as a systemic fungicide. It has both curative and protective actions. It can be absorbed via acropetal translocation. Ramularia, Cercosporaand rhyncosporium can be controlled with this fungicide.

We can use this fungicide in the form of granules and suspensions. Diniconazole can be considered as steroid methylation inhibitor. It has both curative and protective actions. It can be used to regulate ear diseases in several cereal crops (rerefence).

It can use in the form of powder, granules and suspension concentrates. Etridazole can be classified as a contact fungicide. This chemical can have curative and protective actions. Pythium and Phytophthora can infect vegetables and cotton plants.

Etridazole can regulate the fungal growth on these plants. It can be used as a wettable powder and dispersible powder. It does not irritate the eyes and skin.

Fenbuconazole can work as a steroid demethylation inhibitor. It can function as a systemic fungicide. It has eradication, curative and protective functions. It can be used to regulate several diseases that infect vegetables, rice and other crops. It can be used as suspension and emulsifiable concentrates. Hexaconazole can be used as a demethylation inhibitor. We can classify this fungicide as a systemic fungicide. It can regulate pathogens such as basidomycetes and ascomycetes.

These fungal species can act on nuts and fruits. It can be used in the form of granules and suspension concentrates. Penconazole can be considered as sterol demethylation inhibitor. It can be classified as a systemic fungicide. It has both curative and protective actions. It can be used to regulate powdery mildew on the scab and other fruits. Fungal species such as deuteromycetes, basidiomycetes and ascomycetes can be regulated with this chemical. It can be used as wettable powder and emulsions.

Tebuconazole can work as a sterol demethylation inhibitor. It can be classified as asystemic fungicide. It has both curative and protective actions. It can be used in the form of powders, suspensions and gels. Triadimefon can be classified as a demethylation inhibitor. It can be classified as a systemic fungicide. It has both protective and curative properties. It can be used to regulate the powdery mildew on vegetables and fruits. It can be used in the form of powder and emulsifiable concentrates. It can irritate the eye and skin. Triadimenol can be used to disrupt the synthesis of ergosterol and gibberellins. The mitosis process will get damaged. It can be classified as a systemic fungicide. This chemical can have eradicant, curative and protective actions. We can use this chemical in the form of powder, granules and concentrates.

Morpholines can be considered as an effective fungicide. A chemical such as dodemorph can be classified as morpholines. Dodemorph targets the synthesis of ergosterol.

It is classified as a systemic fungicide. It has both curative and protective actions. Most of this chemical can be absorbed via leaves and roots. It can be used as an emulsifiable concentrate. Fenpropimorph can inhibit the biosynthesis of ergosterol.

It can be considered as systemic fungicide and the mode of application is foliar. It exhibits both curative and protective effects. The translocation in the xylem tissue will take acropetal passion.

It can be used as suspension concentrates and emulsions. It can irritate the skin. Tridemorph can inhibit ergosterol biosynthesis. It can be classified as a systemic fungicide. It can be absorbed via roots and leaves. It controls *Odium hevea* and exobasidium species.

The oxathiins can be called as carboxanilides. It can be used as a systemic fungicide. We can use this fungicide to treat the seeds. The bunts and smuts can be regulated with these fungicides.

This chemical can be used to regulate rhizoctonia in several vegetables. It can be used in the form of powder and suspension concentrates. It causes mild irritation to the eyes. Oxycarboxin can be used as a systemic fungicide and we can expect a curative mode of action. We can apply this fungicide on the leaves. It controls rust diseases.

1.7 HERBICIDES

The herbicides are the chemicals that can kill herbs or plants which can create damage to the useful crops. These herbicides can exhibit diversity in their actions.

The herbicides can be classified into different groups. The main groups are inorganic herbicides, bipyridylium herbicides, and organic acid herbicides. These compounds can be derivatives of phenoxy compounds.

Some of the important herbicides are 2,4,5-T, 2,4-D, sodium chlorate, etc. Sodium chloride can be considered as a popular herbicide. Inorganic herbicides such as sodium chlorate can be flammable and act as an oxidizing agent. The bipyridylium herbicides can act on non-selective mode.

Diquat and paraquat can be considered as important bipyridylium herbicides. These compounds can act at the molecular level. They will be disruption in the redox reactions and so some of the reactive oxygen species will be generated.

Due to these reactions, the amount of NADPH will be reduced. These herbicides can damage kidneys. We can expect toxicity in the lung tissues. Diquat does not damage the lungs.

We can observe that this chemical can be absorbed at low levels into the gastrointestinal systems. There was no considerable genotoxicity and mutagenicity. In the case of humans, we can observe both systemic and local effects (reference).

The phenoxy herbicides can be considered as analogs of auxin molecules. They can be used as growth hormones for plants. One of the interesting facts is that these herbicides can produce uncontrolled growth in the weeds. The classification of herbicides is given in figure 1.6.

Group	Mode of Action	Family[2]	Common name	Trade name
1	ACC-ase Inhibitor	'fops'	quiazalotop	Assure II
		'dims'	clethodim	Select
2	ALS inhibitors	sulfonylurea	nicosulfuron	Accent
		imidazolinone	imazethapyr	Pursuit
		–	flumetsulam	
3	Mitosis inhibitors	dinitroalinine	trifluralin	Treflan
4	Synthetic auxins	phenoxy	2,4 D	many
		-	dicamba	Clarity
5	Photosystem II inhibitor[3]	triazine	atrazine	many
6	Photosystem II inhibitor[3]	-	bromoxynil	Buctrill
9	EPSPS inhibitor	-	glyphosate	Roundup
10	Glutamine synthetase inhibitor	-	glufosinate	Ignite
13	Carotene inhibitor	-	clomozone	Command
14	PPO Inhibitors	diphenylether	fomesafen	Reflex
		-	flumioxazin	Valor
15	Lipid inhibitors	acetamide	metolachlor	Dual
		-	dimethenamid-P	Outlook
19	Auxin transportInhibitor	-	diflufenzopyr	Status
22	Photosystem I Inhibitor	bipyridilliums	paraquat	Gramoxone
27	HPPD inhibitor	triketones	termbotrione	Laudis
		-	mesotrione	Callisto

Figure 1.6Classification of herbicides

Source:https://pesticidestewardship.org/wp-content/uploads/sites/4/2016/07/Herbicide-MOA-classification.jpg

They may not induce any considerable effects on humans. Some of the examples of this type are dichlorprop, mecoprop, MCPA and 2,4- D. In the case of 2,4-D there will be salt dissociation and ester hydrolysis. This herbicide does not produce any irritation. We can observe certain anomalies in the liver and adrenal glands. 2,4,5 T usage was stopped in many countries. This is due to the toxicity of this chemical on humans and possible reactions with dioxins. Dicamba exhibits low toxicity and it is acute in nature. Due to the use of this chemical we can expect a proliferation of peroxisomes. Some of the herbicides which can alter the growth of the plants are haloxyfop, fluazifop-P and fluazifop. All these chemicals can be considered as derivatives of phenoxypropionic acid. Some of the aniline derivatives are alachlor, propanil, metolachlor and propachlor. Propanil can be considered as a secondary amine. Due to the usage of these chemicals, there will be lysis of reticulocytes and lower hemoglobin. Alachlor may not be toxic in nature. In the case ofrodents, this chemical can be carcinogenic. It has no specific genotoxic properties.

Monolinuron, linuron and diuron are the chemicals that can be classified as herbicidal ureas. These chemicals can alter the photosynthesis process. These chemicals can exhibit lower toxicity. Bromoxynil and ioxynil can be classified as nitrile herbicides. Some of the broad -spectrum herbicides are simazine, cyanazine and atrazine. The traizoles and triazines can be considered as broad- spectrum herbicides. Glufosinate and glyphosate can be classified as organic phosphorus herbicides. Defoliants and desiccants can be considered crucial in the agriculture field.

1.8 MICROBIAL PESTICIDES

These are the substances that harbor the microbial organisms and they protect plants and their products. These substances can be called as microbial pest control agents or in short MPCAs. Some of the microbes which can be inside these agents are microsporidium, nematode, protozoon, virus, fungi and mostly bacteria. The classification of microbial pesticides is shown in figure 1.7.

Types of Bio-pesticides

Sl. No	Types	Active Ingredients/ Substances it includes	Example
1.	Microbial pesticides	Bacterium, fungus, virus or protozoan.	Strains of *Bacillus thuringiensis* (Bt).
2.	Biochemical pesticides/ herbal pesticides	Naturally occurring substances	Scented plant extracts.
3.	Plant-Incorporated-Protectants (PIPs)	Genetic material/ toxin	Plant producing Bt toxin.

2

Figure 1.7. Classification of Microbial pesticides.

Source:https://www.slideshare.net/bijayauprety/biopesticide-41209874

Ampelomycesquisqualis can be considered as a biopesticide to safeguard grapes but this parasite can produce powdery mildews in other crops. *Coniothyriumminitans* is a fungal member and they can regulate the growth of sclerotinia. This fungal member can survive in any environment. Microbial pesticides can reproduce and multiply.

They can produce secondary metabolites. They offer some benefits than the chemical -based pesticides. *Beauveria bassiana* can provide protection against insects. This organism can produce beauvericin. *Phelbiopsis gigantean* belongs to basidiomycetes. It can grow saprophytically. *Pseudomonas chloraphis* can regulate fungi which mainly attack the root system.

1.9 BIOCIDES

The biocides can be used in many inpowderries. Some of the inpowderrieswhich use the biocides are retort processing systems, pulp mills and oil field water flooding inpowderries. Other biocides include sterilants, disinfectants and sanitizers.

In case of low level disinfection, there will be less bacteriocidal property and a least portion of viruses will get eliminated.

In case of sterilization, autoclaving can be considered as one of the best methods. Some of the biocides which can be used to treat water are MBT or methylene-bis-thiocyanate and GA or glutaraldehyde.

There are many biocides in the market which can be used for this purpose. We can classify these biocides into 2 important classes. They are oxidizers and non-oxidizers.

Some of the biocides which can be used on an inpowderrialscale are phenolics and heavy metals.

The antibiotics used in medicine may develop resistance in bacterial species. This is not true in the case of biocides. We can use some of the biocides for preservation. The classification of the biocides is shown in figure 1.8.

Dibromo nitrilopropionamide or DBNPA can be used on the cells which are either living or dead. Mostly these molecules will adhere to the cell surface. This compound can bind with thiols. It can alter the enzymes and other proteins.

Class and biocide(s)	Active constituent
Alcohols	
Ethanol	Ethanol
Isopropanol	Isopropanol
Biguanide	
Hibitane	Chlorhexidine-digluconate
Halogens	
Chloramine-T	Sodium-tosyl-chloramide
Isobetadine[a]	Povidone-iodine
Sodium hypochlorite	Hypochlorite
Peroxygens	
Hydrogen peroxide	Hydrogen peroxide
Peracetic acid	Peracetic acid
Phenol	
Dettol	Chloroxylenol
QACs	
Cedium[a]	Benzalkonium chloride
Cetrimide	Cetrimide
Other	
Furacine[a]	Nitrofurazone

Figure 1.8. Classification of Biocides.

Source:https://www.researchgate.net/publication/42975644_Inhibitory_Effect_of_Biocides_on_the_Viable_Masses_and_Matrices_of_Staphylococcus_aureus_and_Pseudomonas_aeruginosa_Biofilms/figures?lo=1

It can kill both gram positive bacteria and gram negative bacteria. This chemical can be used in pulp processing and cooling towers as its lifetime is much less atleast in the lower pH conditions. It can produce oral toxicity and in some cases, there will be peroral toxicity. It can irritate skin and in some cases eyes. Methylenebisthiocyanate or MBT can damage the metabolism of the cell. This compound can bind with the iron molecules present in the enzymes. We can say that this compound can interact and disrupt cytochromes. The energy transfer process can also be interrupted. We can use MBT in the form of dilute fluid. Mostly we can use it in the form of dry powder. For three decades, MBT is available in the open market. It can kill algae, bacteria, molds and yeast. It can act on anaerobic sulfate reducing bacterial species. Quarternary ammonium compounds known as quats. They are basically cationic compounds which comprise of N-heterocyclic groups. They can be most sensitive in the salt waters. If the organic load is higher, then these compounds can be more sensitive. These compounds kill gram positive bacteria. ADAC or Benzyl dimethyl ammonium chloride and DDAC or Didecyl dimethyl ammonium chloride can be considered as commonly used quats. These compounds release active species and these chemical species can kill several microbes.

Biocides exhibit biostatic properties. Most of the biocides can be used as preservatives. We can use these biocides in the form of acid formulations. Iso-thiazolones can be considered as a bacteriostatic. It will consume a lot of time to get the desired activity. The Krebs cycle will get disrupted due to the activity of this chemical, The cells will get damaged as the thiols get eliminated. Tetra hydroxymethyl phosphonium sulfate can be called as THPS. This chemical can be considered as a quarternary phosphonium compound. This compound can be used in the form of aqueous solutions. Some of these compounds can be considered as surfactants. They can kill gram positive and gram negative bacteria. They can alter the metabolisms of algae, fungi and other S-reducing bacteria.

Paracetic acid or PAA can be used as a protein denaturation agent. It will break sulfur bonds which results in the inactivation of the enzymes. The membrane transport mechanisms can be disrupted due to this chemical. It can control bacteria and can work better than peroxide. Glutaraldehyde or GA can be classified as aliphatic dialdehyde. It can react with thiols and amines. In acidic pH, this chemical can be active.

2

Herbicides in Crop Production

CONTENTS

2.1 INTRODUCTION

The weeds can be regulated with the aid of herbicides. Weeds are the plants that can damage the growth of the crop plants. Due to weeds, the economy of several countries dwindle. Weeds can interfere with ecosystems. They can spoil the water resources. Some of the weeds can be toxic to both animals and humans. These plants can harbor diseases and a variety of pests. They can even damage crop harvesting machines. Most of the weeds can compete with crop plants for nutrients, moisture and light. We can use herbicides to regulate the growth of weeds. The weeds were removed by hand in 10000 B.C. and in 1947; these weeds were controlled by chemicals. The chemical used to control these weeds include sulfuric acid, copper salts, rock salts and oil wastes. Dinitrio compound named as DNOC can be considered as a contact herbicide. It was first used in France in the year 1993. The plants treated with the herbicides are shown in figure 2.1.

Figure 2.1. Treatment with herbicides (right-untreated: Left-treated).

Source: http://www.plantphysiol.org/content/166/3/1119

2.2 CLASSIFICATION

The herbicides can be classified according to the mode of application, timing and activity. The herbicides based on their translocation are classified into systemic and non-systemic herbicides. The systemic herbicides can be translocated via the vascular system. They can be translocated with nutrients and water. These herbicides can act on perennial weeds. The action of these herbicides can be slow. Dicamba can be considered as systemic herbicide and it acts on selected crops. Glufosinate and glyphosate can be considered as systemic herbicides but they are non-selective in nature. The non-systemic herbicide can act on a specific portion of a weed plant. We cannot absorb translocation with these herbicides. The perennial plants are least affected due to these herbicides. These herbicides can act faster than systemic herbicides. Some of the compounds which are considered as non-systemic are bentazon and bromoxynil.

The evolution of the herbicides is represented in figure 2.2.

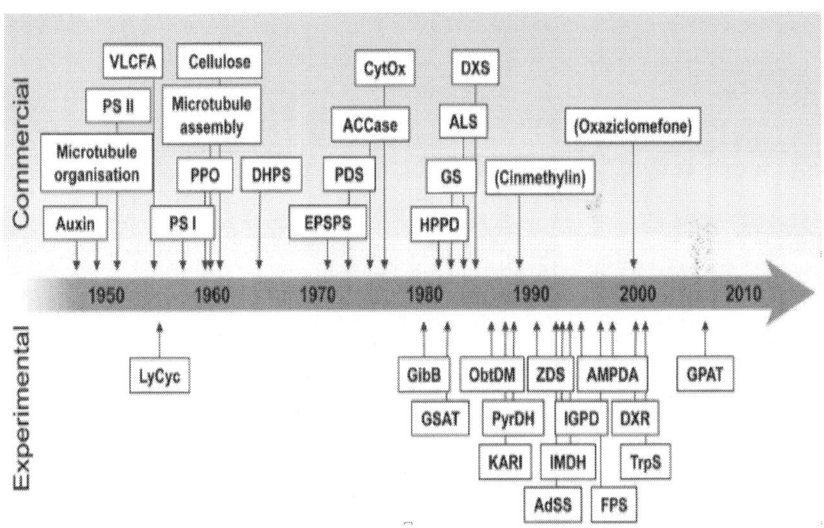

Figure 2.2. Evolution of herbicides.

Source:http://www.plantphysiol.org/content/166/3/1119

We can classify herbicides based on their time of application. They are classified into preplant and preemergence herbicides. The preplant herbicides must be applied to the soils. They can be considered as non-selective herbicides. Once applied to the soil, these herbicides will get incorporated and they will not be decomposed due to light. They will not be susceptible to volatility. They can control weeds effectively. The herbicides which are volatile in nature can be applied before sowing. Dazomet and metam sodium can be considered as preplant herbicides. The crops which are being protected with the preplant herbicides are strawberries, corn and tomatoes. If the herbicides are applied prior to the emergence of plant seedling, we call such herbicides as preemergence herbicides.

The herbicides which can be used after the emergence of the seedlings can be called postemergence herbicides. These herbicides can be used as systemic, contact, non- selective and selective. They can be absorbed in some plants. These herbicides may not be used whilst rain as it will inactive the herbicide. These herbicides can be used in the form of liquid. 2, 4 D can be considered as a post-emergence herbicide. It can be absorbed through leaves. It can be classified as a systemic and selective herbicide.

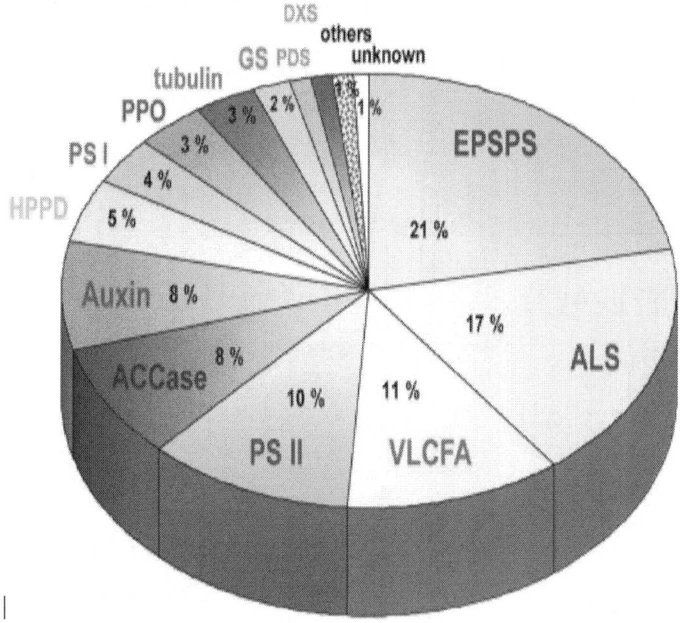

Figure 2.3. Herbicides-Market shares (2010).

Source:http://www.plantphysiol.org/content/166/3/1119

The market share of the herbicides in 2010 is represented in figure 2.3. The herbicides are classified into soil- applied herbicides and foliar -applied herbicides. The herbicides which can be applied to the soils can be called soil -applied herbicides. These herbicides can be absorbed by the shoots and roots of the seedlings emerging from the soil. We can use this type of herbicides for both preemergence and preplant treatment. The weeds can absorb these pesticides through specialized mechanisms. If these herbicides adhere to the soil particles and organic matter, there will be a minimum amount of herbicide concentration available for weeds to absorb. Trifluralin and EPTC can be considered as soil -applied herbicides. The foliar -applied pesticides can be applied to a specific portion of plants. Mostly these herbicides can affect the tissues which are exposed. These herbicides can be translocated to specific sites of a plant. They can serve as both contact herbicides and systemic herbicides. The cell wall, wax, cuticle are susceptible to the absorption of these herbicides, Dicamba, 2,4 D and Glyphosate can be used as foliar- applied herbicide.

We can classify these herbicides based on their specificity. They can be classified as selective herbicides and non-selective herbicides. The selective herbicides can influence weeds and they will not affect the crop plants. They can act based on the physiology of the weeds. Some of the other properties which must be considered while selecting this type of herbicides are morphological properties, absorption and translocation. Dicamba, mecroprop and 2,4 D can regulate the broadleaf weeds. These herbicides may not affect the turfgrasses. The non-selective herbicides can kill any plant material. We can use non-selective herbicides to clear weeds grown in the rail tracks, waste grounds and inpowderrial areas. Glyphosate, glufosinate and paraquat can be considered as non-selective herbicides.

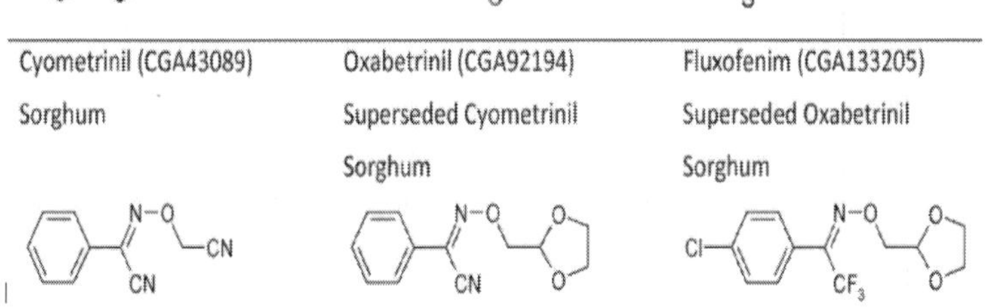

Figure 2.4. Seed treatment.

Source: http://www.plantphysiol.org/content/166/3/1119

The herbicides used in seed treatment can be represented in figure 2.4. We can classify the herbicides depending on their site of action. Group I(A) herbicide comprise of aryloxyphenoxy propionate. It can act as ACCase or acetyl CoA carboxylase inhibitor. Cyclohexanedione can exhibit similar properties. Group 2 (B) herbicides can influence ALS or acetolactate synthase. Some of the chemicals which show this property are imidazolinone, pyrimidinylthobenzoate, sulfonylamino-carbonyltriazolinone, sulfonyl urea and triazolopyrimidine. Group 3(K1) can be used to inhibit the microtubule organization or assembly. Dinitroaniline and pyridine fall under this group. 4(O) can be considered as synthetic auxins. Phenoxy compounds can act as herbicides. Some of the other chemicals which come under this group are benzoic acid, carboxylic acid,Quinalinecarboxcyclic acid fall under this group. Group 5(Cl) can influence photosystem II especially site A. The chemicals such as phenyl carbamate, pyridazinone, triazines, triazinone and uracil come under this group. Group 6(C3) can influence site B of photosystem II.

Benzothiadiazole, Nitrile and phenyl-pyridazine come under this group. Group 7 (C2) can inhibit site A of photosystem II. Amides and urea come under this group. Group 8(N) can affect the synthesis of lipids. Thiocarbamate comes under this group.

Group 9(G) can affect enolpyruvyl shikimate phosphate synthase. Group 10 (H) can affect glutamine synthetase. Group 11(F3) can influence the biosynthesis of carotenoids. Triazole can be classified as 11(F3) herbicides. Group 12 (F1) can influence phytoene desaturase. Pyridinecarboxamide and pyridazinone come under this group.

Glyphosate can be used as broad -spectrum herbicide. This chemical targets EPSPS or enolpyruvyl shikimate-3- phosphate synthase. The history of herbicides is represented in figure 2.5.

MoA, Target Site	Chemical Family	Examples	Use	Earliest Reports
Unspecific	Inorganic herbicides	H_2SO_4, Cu_2SO_4, $FeSO_4$, $NaAsO_2$	Total	1874
Uncouplers	Dinitrophenoles	dinitro-*ortho*-cresol	Post, dicots	1934
Auxins	Aryloxyalkanoic acid derivatives	2,4-Dichlorophenoxyacetic acid	Post, dicots in cereals	1942
Microtubule organization	Arylcarbamates	Propham, chloropropham	Pre, monocots in various crops	1946
Lipid synthesis	Chloroaliphatic acids	TCA, dalapon	Pre, monocots in various crops	1947
	Thiocarbamates	EPTC, triallate	Pre, monocots and dicots in various crops	1954
PSII	Arylureas	Monuron, diuron, isoproturon, linuron	Pre and Post, monocots and dicots in various crops	1951
	1,3,5-Triazines	Atrazine, simazine	Pre and Post, broad spectrum in corn	1952
	Pyridazines	Chloridazon	Pre, dicots in sugar beet	1962
	Uracils	Bromacil, terbacil, lenacil	Soil applied, broad spectrum in various crops	1963
	Biscarbamates	Phenmedipham	Post, dicots in sugar beet	1968
	1,2,4-Triazinones	Metribuzin	Pre in soybean	1971
Very-long-chain fatty acid biosynthesis	Chloroacetamides	Allidochlor, alachlor	Pre, monocots and dicots	1956
PSI	Bipyridyliums	Diquat, paraquat	Nonselective	1958
Protoporphyrinogen oxidase	Diphenyl ethers	Nitrofen, acifluorfen	Pre and Post, various crops	1960
	Oxadiazoles	Oxadiazon	Rice, nonselective	1969
Microtubule assembly	Dinitroanilines	Trifluralin, pendimethalin	Pre against monocots and dicots	1960
Cellulose biosynthesis	Nitriles	Dichlobenil	Plantations	1960
5-*Enol*pyruvylshikimate 3-phosphate synthase	Glys	Glyphosate	Post, nonselective	1971
Phytoene desaturase	Pyridazinones	Norflurazon	Pre and Post in cotton	1973
ACCase	Aryloxyphenoxy propanoates	Diclofop, fluazifop	Post, grasses	1975
	Cyclohexane diones	Alloxydim, sethoxydim	Post, grasses	1976
Gln synthetase		Glufosinate	Nonselective	1981
AHAS or ALS	Sulfonylureas	Chlorsulfuron, metsulfuron	Monocots and dicots in various crops	1982

MoA, Target Site	Chemical Family	Examples	Use	Earliest Reports
	Imidazolinones	Imazapyr, imazethapyr	Nonselective or selective in soybean	1983
	Pyrimidinyl benzoates	Bispyribac sodium	Rice	1994
HPPD		Pyrazolynate, sulcotrione	Various crops, monocots and dicots	1984

Figure 2.5. History of herbicides.

Source:http://www.plantphysiol.org/content/166/3/1119

2.3 TOXICOLOGY

Common salt or sodium chloride can be considered as inorganic herbicides. Bipyridyliu, herbicides can be used as a non-selective herbicide. Some of the popular herbicides are diquat and paraquat. The pesticides which can influence plants such as weeds are termed as herbicides. It is not much surprising to mention that nematicides, fungicides and insecticides have less influence on plants than herbicides. These herbicides will interrupt with the biochemical reactions.

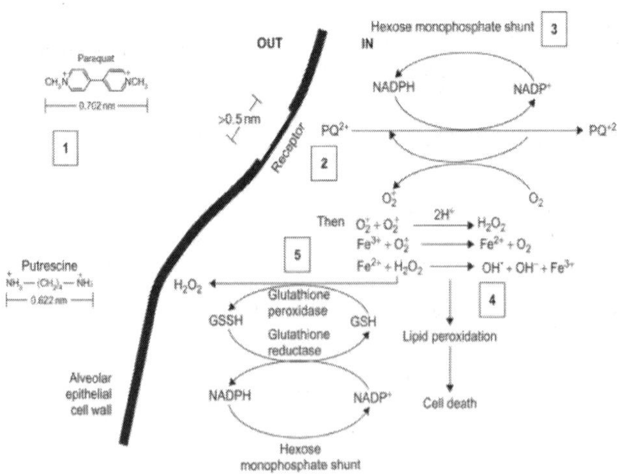

Figure 2.6. Toxicology of herbicides.

Source: http://www.plantphysiol.org/content/166/3/1119

The toxicology of herbicides is shown in figure 2.6. The action of herbicides can be classified into 3 main categories. The herbicides can inhibit the growth and development of plants. Some herbicides can damage cell membranes. Some herbicides can influence the biosynthesis pathways inplants. Some of the pathways which can be altered due to the activity of herbicides are amino acid synthesis, lipid biosynthesis and carotenoid biosynthesis. The herbicides can affect the sites where there will be the production of amino acids. The branched and aromatic amino acids can be disrupted. There will be altered protoporphyrin IX, excess storage of chlorophyll precursors, electron capture in PS I and lipid peroxidation. When treated with the herbicides, the plants can exhibit white leaves due to discoloration. We can define it as albino growth in most of the cases. As the chlorophyll content is decreased, the plants cannot perform photosynthesis and so they eventually die. There will be an onset of necrosis as peroxidative reactions occur. There will be the development of chlorosis.

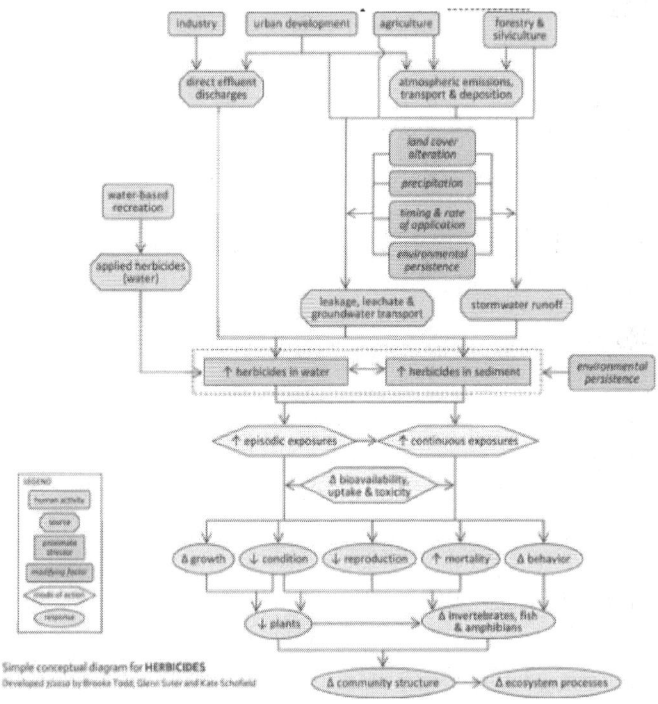

Figure 2.7. Conceptual diagram of herbicides.

Source:https://www.epa.gov/sites/production/files/2015-11/herbicides-cd_sim_500.jpg

The conceptual diagram of the herbicides is given in figure 2.7.Due to photooxidation, there will be loss of chlorophyll. The carotenoids will protect the chlorophyll from getting oxidized. There will be destruction of chlorophyll and membrane destruction.

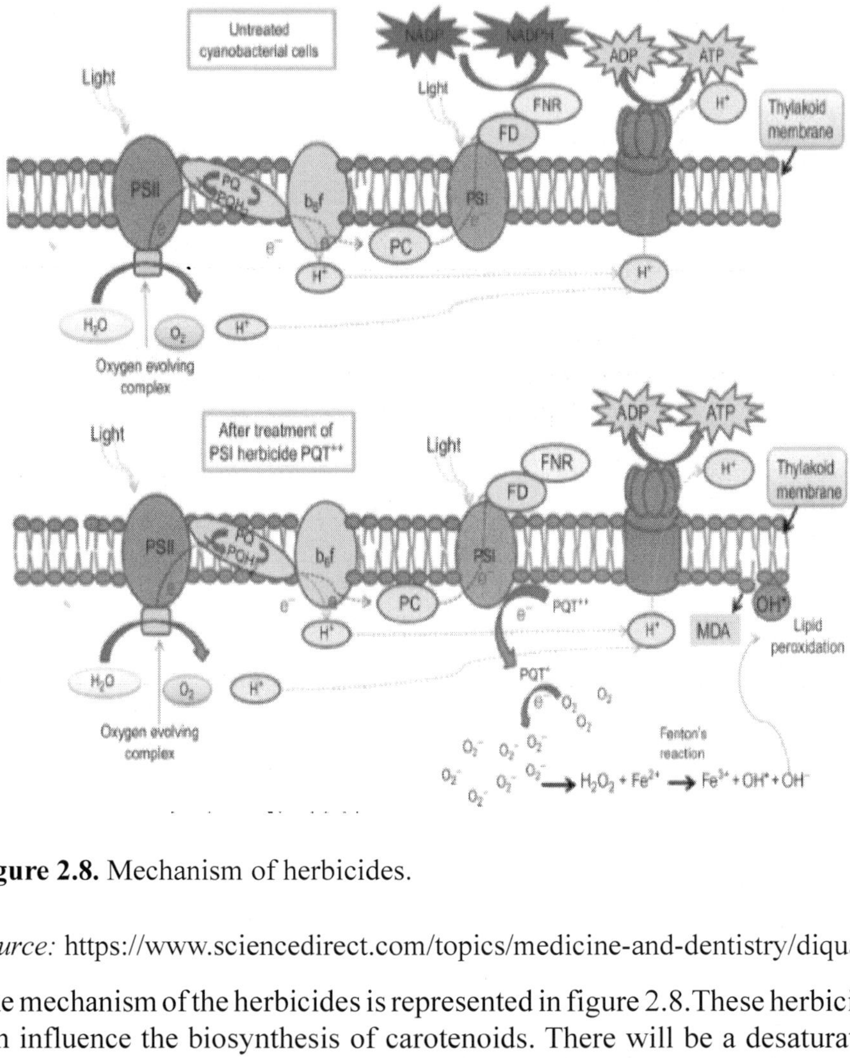

Figure 2.8. Mechanism of herbicides.

Source: https://www.sciencedirect.com/topics/medicine-and-dentistry/diquat

The mechanism of the herbicides is represented in figure 2.8.These herbicides can influence the biosynthesis of carotenoids. There will be a desaturation of phytoene. Mthoxyphenone and dichlormate can act on these sites. When the herbicides act on a particular site, there will be inhibition of gibberellic acids, chlorophylls and carotenoids. It is crucial to understand that plants require lipids in higher amounts. As the plant grows, there will be a demand for new lipids to form membranes. Because of the action of herbicides, the lipid production will be hindered and so the meristematic tissues. The cyclohexanediones and aryloxyphenoxypropionates can be considered as popular herbicide classes. These compounds can alter the biosynthesis of lipids. Due to the effect of these compounds, there will be abnormal growth of the grana. There will be vesiculation in the cytoplasm. The cells can suffer from senescence. Glyphosate can stop the biosynthesis of aromatic amino acids. The weeds can be killed gradually. It will take atleast 15days so that the weeds can die. There will be the development of necrosis and chlorosis post- treatment. The tyrosine and phenylalanine levels will be reduced due to the application of this herbicide.

There are three types of herbicides that can stop the amino acid biosynthesis. The branched amino acids will get affected much. Some of the herbicides which can influence the biosynthesis of amino acids are triazolopyrimidine, slfonylureas and imidazolinones. Imazaquin, chlorsulfuron and flumetsulamare considered as important herbicides. We can expect a gradual and slower effect on the weeds. After some weeks, the activity of these herbicides will be visible. The first tissue which can be affected is the meristematic tissue. These herbicides can inhibit cell division. The Gl and G2 phases can be affected due to these herbicides. The plant cell can be completely functional if the apoplast, plasma membrane and other membrane systems are intact. The herbicides can produce free radical reactions and this is due to peroxidation of the lipids. We can observe three main stages in the peroxidation of the lipids. They are initiation followed by propagation and finally termination. The chlorophyll, OH radical and oxygen radical can be considered as initiation factors. There will be the formation of peroxidized lipid radicals. These radicals can be reduced so that there will be the formation of lipid peroxides. This stage can be called a propagation reaction. Slowly there will be degradation of the lipid peroxides which can yield hydrocarbons with short chains. These reactions can destabilize the lipids present in the membranes. These mechanisms can stop the PS II electron flow. The photosynthesis process will be inhibited.

The uracils, ureas and triazines can stop the electron flow in the photosystem II. These herbicides will influence the light reactions in the photosynthesis process. Terbacil and bromacil are the uracil based herbicides. Monuron and diuron can be classified as urea- based herbicides. Cyanazine and atrazine can be considered as triazines based chemicals. The enzymes, proteins and pigments which are needed for photosynthesis are concentrated in the chloroplast molecules. These herbicides can influence the photosynthesis process by associating with quinine binding proteins. It is a 32-kDa protein. Due to the impairment of the photosynthesis process, there will be problems with the production of carbohydrates and carbon fixation. We can observe chlorosis in the leaves due to the application of these pesticides.

There will be the production of lipid radicals from the reactions of these herbicides on the PUFA or polyunsaturated fatty acids. The chlorophyll which is in its triplet form can react with the oxygen molecules to yield singlet oxygen or free radicals of oxygen. These oxygen molecules can react with the lipid radicals formed so that there will be a peroxidation process. These free radicals can do damage to the membranes of the weeds.

Paraquat and diquat can accept electrons which are derived from the photosystem I or PS I. This photosystem is observed in the light reactions. The activity of these compounds will be in the ferredoxin areas. These compounds may not cause much damage to the weeds. Application of diphenyl ether associated herbicides such as oxyfluorfen, lactofen and acifluorfen can cause necrosis in the leaves. We can observe water -soaked regions in the weed plants due to the application of these herbicides. The symptom such as water soaking can indicate that the membrane is damaged and the herbicide applied is effective. Most of the cases, light can activate the action of herbicides. Glufosinate can inhibit the enzyme glutamine synthetase. We can call this herbicide as PPT or phosphinothricin. It is one of the synthetic herbicides. Glufosinateisan amalgamation of d and l isomers of PPT. It is evident that L-isomer will be active in many cases. Another chemical that has PPT as the main ingredient is bilanafos. It is a tripeptide. This chemical is not categorized as an herbicide but it can be transformed to aherbicide due to hydrolysis reactions in the plants. The herbicide glufosinate can react on all plants and hence it is considered as non-selective. We can observe both necrosis and chlorosis post- application of glufosinate. Mostly this herbicide can take effect after 2 days.

Post application of this herbicide, the level of ammonia will get increased rapidly. It was observed that there is atleast 4 times increase in the ammonia levels. The levels of ammonia can be measured with the decreased quantities of glycine, serine and aspartate. The glufosinate activity can be understood by the glutamine levels. Most of these herbicides can inhibit growth. The action of herbicides on the plants is possible even if these chemicals do not cause death. This is because of the fact that the seedlings require much of the carbon and carbohydrates. The production of these carbohydrates can be hindered due to the activity of these herbicides. The herbicides can inhibit cell division. There may not be observe any kind of mitotic stages, especially in the meristematic zones. This is evident when observed through a light microscope. The herbicides may not affect DNA synthesis. It is evident through several experiments that these herbicides can halt G1 and G2 stages. Some of the herbicides which can act as growth inhibitors are acetochlor, metolachlor and alachlor. These chemicals can alter G1 or in some cases G2.

The meristematic cells will divide into daughter cells due to mitotic divisions. The herbicides can disturb the cell division. They will damage the sequence in which the cell division occurs. The herbicides such as dinitroaniline derivatives can disturb the formation of the spindle. Some of these herbicides are prodiamine, benefin, oryzalin and trifluralin. These herbicides can disrupt the process of tubulin polymerization. The chromosomal movements will be disrupted due to the action of these herbicides.

We can expect to swell in the apical zones of the seedling. This symptom is due to the application of the herbicides. Slowly the meristems will disappear. There will be swelling in the root tip regions on the application of carbamate and dinitroaniline herbicides. Isoxaben and dichlobenil are herbicides which can lower the formation of the cell wall.

The fungal growth can be prevented by disrupting the formation of the cell wall. The cell wall growth can be impaired by isoxaben and dichlobenil. The herbicides can influence the microtubules and due to this, there will be improper enlargement and division of cells. Dichlobenil can disrupt the biosynthesis of the cell wall. There will be an impairment of cell wall synthesis. Isoxaben may not reduce acetate, uracil and leucine. These herbicides may not require light. Auxins can influence the plant development both directly and indirectly. The auxin concentrations can be regulated by the cell. There are some synthetic auxins that can regulate plant development. Some of the synthetic auxins are MCPA, dicamba and 2,4-D. These biochemical changes caused due to these auxins can stimulate protein synthesis and nucleic acids.

2.4 PLANT PHYSIOLOGY

The strength of the herbicides can be based on the active portion and proportion of the herbicides which can reach the region of activity. There are several barriers that can hinder the migration of these compounds in the plant. Mostly when these herbicides are applied, they need to pass through several barriers before they reach the sites such xylem or any other tissues.

The waxy leaf surface can protect the leaf surface from the effects of the environment. The structure of the compounds can be complex and diverse in nature. In general, the thickness of the cuticle will range from 0.1 to 13 micrometers in thickness. We can observe 3 main components in the cuticle. They are cutin matrix, cuticular and epicuticular waxes. We may not observe a homogeneous mode of arrangement.

If waxes are present on the surface, then we can call them epicuticular waxes. We can remove this wax by immersing the leaves in the organic solvents. These waxes comprise of flavinoids, sterols, esters, ketones and aldehydes with several other compounds. These waxes are predominant in the brassica and leek leaves.

There are some crystals which are present in these waxes. They are in the form of aggregates, crusts, granules and rods. The adaxial surface and abaxial surface can be having different proportions of these crystals. Because of these crystals, these leaves will appear dull.

In case these crystals are absent, we can see that these leaves are shiny in their appearance. The hydroxylated fatty acids can result in the formation of cutin. The cuticular waxes are present in the epidermal cells.

The biosynthesis of the cutin molecules can be based on acyltransferase and oxidoreductase genes. Enzymes involved in the production of the cutin molecules are acyltransferases, acyl-CoA synthase and fatty acid oxidases. These enzymes belong to GPAT, LACS and CYP86A sub-family. These enzymes can be observed in the endoplasmic.

There are some pores that can allow these herbicides to translocate from one place to another place until they reach the site of action. There will be some proteins that can help in the translocation of the lipid molecules. These proteins can be called as LTPs or lipid transfer proteins. These proteins can transfer the lipids via endoplasmic reticulum.

The biosynthesis of the lipids present in the cuticle can react with the environmental conditions. In the case of disparities in the soil conditions, photoperiods and light intensities, they will react and it is measurable. The cuticle will get thickened whilst unfavorable conditions.

The prevalent adverse conditions are low water availability, photon flux and lower temperatures. The absorption of herbicides will be higher only if the plant species are grown under favorable conditions.

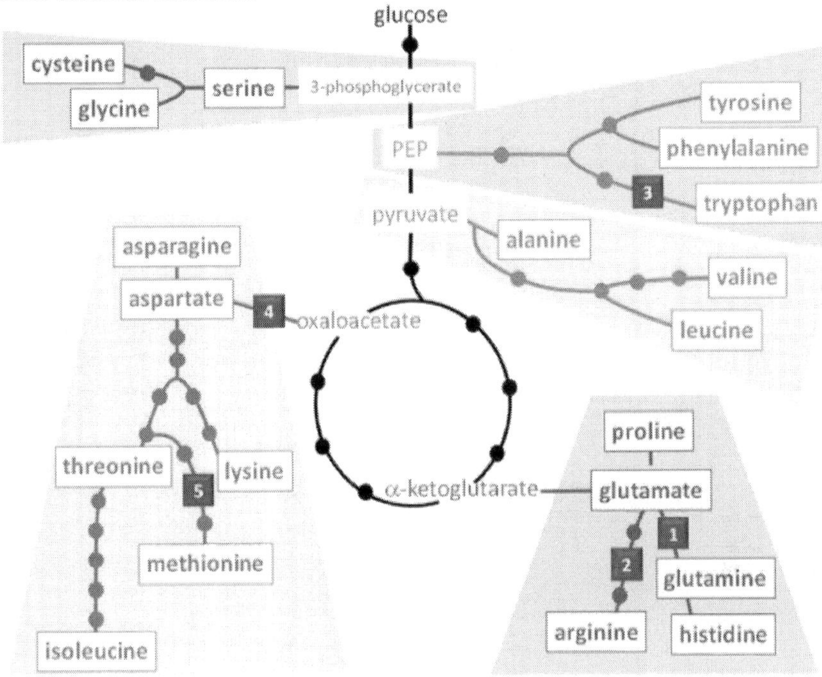

Figure 2.9. Biosynthesis pathways and sites for herbicides.

Source: http://www.plantphysiol.org/content/166/3/1090

The biosynthetic pathways and the sites of herbicides are represented in figure 2.9. These chemicals can get absorbed through the leaves only if an intermediate compound named as an adjuvant is added to the mixture.

The surfactants can be added to these herbicides so that their deposition rates on the leaves can be maximized. Mostly these surfactants arein the form of amphipathic molecules.

The surfactants can be classified into four main classes. They are amphoteric, non-ionic, cationic and anionic surfactants.

The amphoteric surfactants are the molecules that can comprise hydrophilic groups. They can convert into anionic surfactants if pH increases.

They can also get converted into cationic surfactants if the pH values is dropped. The non-ionic surfactants can have no charge. The cationic surfactants are positively charged ions.

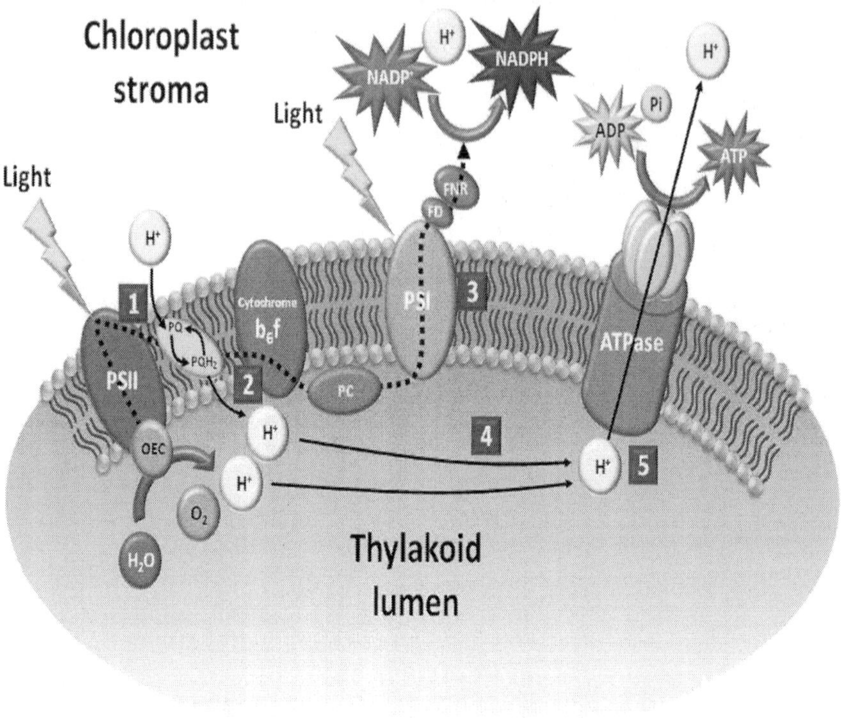

Figure 2.10. Target sites for herbicides.

Source: http://www.plantphysiol.org/content/166/3/1090

The target sites of the herbicides are represented in figure 2.10. The hydrophilic group is balanced by the positively charged groups. Quaternary ammonium can be considered as cationic surfactants.

The anionic surfactants can have negatively charged ions. Here the hydrophobic groups are balanced by negatively charged groups. Carboxyl groups can be considered as anionic surfactants.

Hydrophilic lipophilic balance or HLB is the ratio between the hydrophilic and lipophilic groups. The substances with lower HLB value can be soluble in water.

We can use them as activators and modifiers. These compounds can lower the surface tension of the droplets. The biosynthesis of chlorophyll and the carotenoids is represented in figure 2.11.

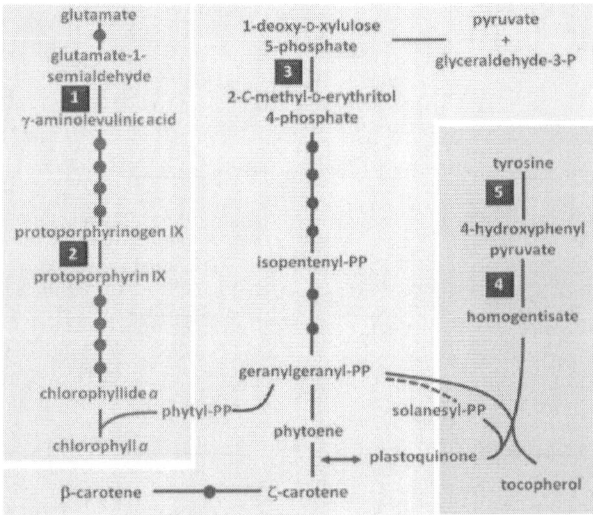

Figure 2.11. Biosynthesis of chlorophyll and carotenoids.

Source: http://www.plantphysiol.org/content/166/3/1090

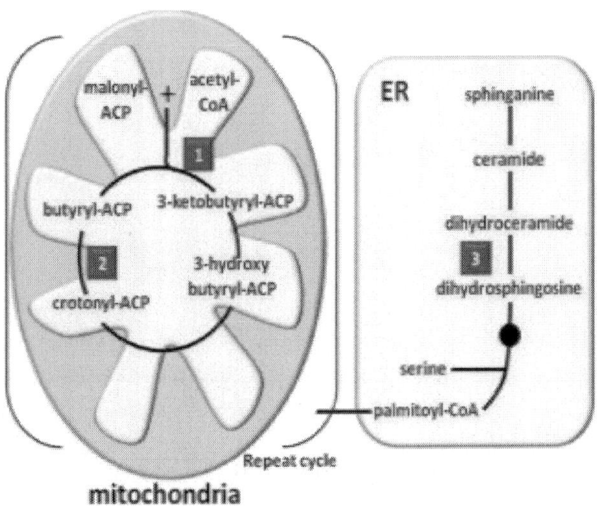

Figure 2.12. Lipid biosynthesis and herbicide sites.

Source: http://www.plantphysiol.org/content/166/3/1090

The biosynthesis of lipids along with the sites for herbicidal action is represented in figure 2.12.

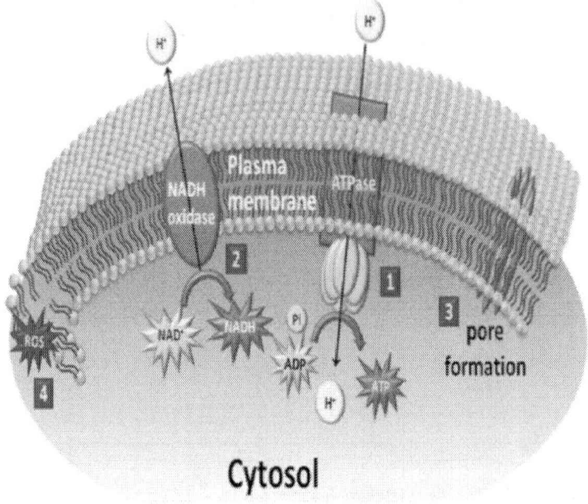

Figure 2.13. Lipid stability.

Source: http://www.plantphysiol.org/content/166/3/1090

The stability of the lipids and the activity of herbicides on the stability are explained in figure 2.13.

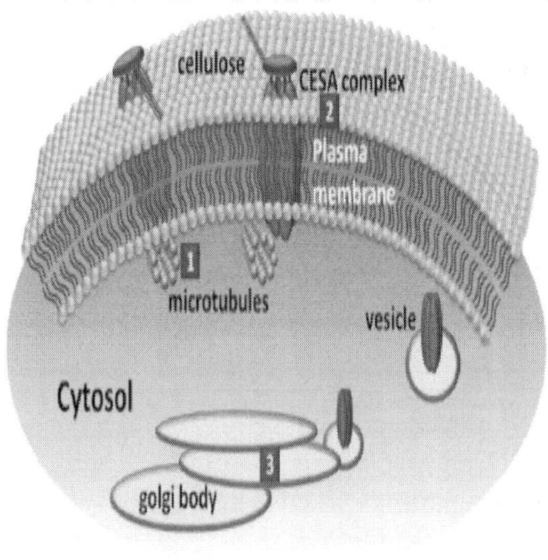

Figure 2.14. Microsomes and herbicides.

Source: http://www.plantphysiol.org/content/166/3/1090

The action of herbicides on microsomes is shown in figure 2.14.

2.5 PREPARATION

The herbicides may not kill the weeds completely. The concentration of the herbicides is crucial in determining their activity. When the herbicides are mixed with water or solid, we call it as formulation. It is a mixture of formulants and ingredients. The formulations can be solid or liquid in nature. If a herbicide is mixed with an organic solvent, then it is called as an emulsifiable concentrate. Powder can yield a good amount of surface coverage. If herbicide is injected into polymeric particles, we call such a type of herbicide formulations as microcapsules. The concentrated aqueous dispersions is called as suspension concentrates. The aqueous concentrates require water, surfactant, wetter and active ingredient. They are easily mixed. They exhibit low volatility and phytotoxicity. They are cheap. These concentrates are frosts sensitive and corrode metal. Powder can comprise a carrier and active ingredient. They can be easily handled. We can easily apply and are cheap to produce. Mostly their flowability can be influenced by moisture. The user can contaminate this type of formulations. The emulsifiable concentrate comprisesof activator, emulsifier, solvent and active ingredient. They exhibit a high rate of efficacy. They can be easily produced and handled. They can be volatile and corrode containers. They are sensitive to frost. The microcapsules contain a preservative, anti-foaming agent, thickener and emulsifier. It also contains a solvent and active ingredients. They can be produced easily. They require less solvent.

Formulations	Mixing/Loading Hazards	Phytotoxicity	Effect on Application Equipment	Agitation Required	Method of Application	Compatible With Other Formulations
wettable powders	dust inhalation	safe	abrasive	yes	foliar and soil	highly
dry flowables/ water-dispersible powders	safe	safe	abrasive	yes	foliar and soil	good
soluble powders	dust inhalation	usually safe	non-abrasive	no	foliar and soil	fair
emulsifiable concentrates	spills and splashes	possible	may affect rubber parts	yes	foliar and soil	fair
liquid flowables	spills and splashes	possible	may affect rubber parts	yes	foliar and soil	fair
solutions	spills and splashes	safe	non-abrasive	no	foliar and soil	fair
granules and pellets	safe	safe	-	no	soil	-

Figure 2.15. Formulation of herbicides.

Source: https://erams.com/static/wqtool/PDFs/bmps_colorado/00558.pdf

The packaging can be expensive. They are sensitive to frost. In the case of high temperatures, they can get thickened. The suspension concentrates contains a preservative, eater, anti-foaming and anti-freezing agent, dispersant, wetting agent and active ingredients. They do not require solvents. They have high ingredient concentrations. They can be easily mixed. They can produce phytotoxicity but are frost -sensitive. If stored for longer periods, they can settle out. The ultra -low volume surfactants comprise of a viscosity modifier, oil and active ingredients. They are cheap and used readily. We can apply them even in low concentrations. They require specialized equipment. They are highly toxic. The labor required for this operation can be high. The water-dispersible granules contain wetting agents, carrier and active ingredients. They don't require a solvent. They can be tolerant of frost. They can be handled easily and needs low powdering. The main problem with them is that they are expensive. The wettablepowder contains a wetting agent, carrier and active ingredient. They don't need a solvent. They are tolerant of frost. They can be easily handled. They are cheap. They can produce powder and hard to mix.

The absorption of the herbicides will be mainly from the roots. They can get translocated through xylem and phloem. The absorption can be through the water or air phase. Bifenox and triallate can migrate in the moist soils in their vapor phase. Simazine and monuron can be diffused in their aqueous phase. The herbicides which are lipophilic can be absorbed by the roots as they have higher vapor pressures. The translocation process can be done via apoplastic and symplastic pathways. The apoplasticpathway allows the substances to move as per the concentration gradient. Symplastic pathway permits the translocation process through plasmodesmata. There should be cytoplasmic continuity in the cortical cells of roots. The herbicide metabolism can be crucial in its movement through the vessels. The mobility of the parent molecule can be based on the hydroxylation of the aryl groups.

2.6 SELECTIVITY

The phytotoxicity can be considered as important property that renders the herbicidal application useful. Due to the application of the herbicides, there will be competitive effects that will kill only weeds and keep the crop unprotected. The selectivity is based on the dose of the herbicide. The rate of application can define the extent of crop tolerance. The susceptibility of the weeds to herbicides can be determined by their selectivity. The concentration of the herbicides which can reach the site of action can be crucial in defining the phytotoxicity of that herbicide.

Family	Herbicides	Mode of Action	Application Timing	Selectivity[1] Grasses	Broadleaf
Dinitroanalines	Treflan/Prowl/ Sonalan	growth inhibitor	mostly pre- and preplant emergence	yes	few
Thiocarbamates	Sutan+/Eptam/ Eradicane	growth inhibitor	mostly preplant emergence	seedlings	a few annuals
Chloroacetamides	Lasso/Dual/ Frontier/Harness/ Surpass/Ramrod	growth inhibitor	mostly pre- and preplant emergence	yes	a few
Triazines	Aatrex/Bladex/ Evik/Milogard/ Igran/Sencor/Lexone	photosynthesis inhibitor	mostly post- and pre-emergence	some	yes
Substituted ureas	Lorox	photosynthesis inhibitor	mostly post- and pre-emergence	many	yes
Organic P compounds	Roundup/ Touchdown	amino acid inhibitor	mostly postemergence	nonselective	nonselective
Dipyridiniums	Cyclone/ Gramoxone Extra	cell disruptor	mostly postemergence	nonselective	nonselective
Sulfonylureas	Accent/Beacon/	amino acid inhibitor	mostly postemergence	yes	yes
	Permit/Ally/ Harmony/Pinnacle/ Peak/Glean/Amber/ Basis/Battalion				
Imidazolinones	Pursuit/Assert	amino acid inhibitor	mostly postemergence	many	yes
Benzonitriles	Buctril	cell disruptor	mostly postemergence	no	yes
Phenoxys	2,4-D/MCPA	growth regulator	mostly postemergence	no	most
Benzoic Acids	Banvel	growth regulator	mostly postemergence	no	most
Picolinic Acids	Tordon/Stinger	growth regulator	mostly postemergence	no	most

[1]Value indicated in these column are used for generalizations of herbicidal activity with respect to chemical family.

Figure 2.16. Selectivity of herbicides.

Source: https://erams.com/static/wqtool/PDFs/bmps_colorado/00563.pdf

The selectivity of the herbicides is represented in figure 2.16. The adsorption variations can be used to determine the type of herbicide to be used. The transport of herbicides in the plant body can be for short distance or longer and it will be based on the plant. The accumulation of the herbicides can be based on translocation methods and patterns. The selectivity can be based on multiple factors that can bring about complex interactions. We can observe differences in plant absorption capabilities. Most of the plants are based on biochemical reactions and mechanisms. The plants are equipped with defensemechanisms that can protect them against foreign bodies or chemicals. These foreign chemicals are called as xenobiotics. The enzymes which can attack these xenobiotics have a wide range of activity. Flavonoids can be considered as compounds responsible to defend the plant from the xenobiotics. The enzymes which can protect the plants from these foreign compounds can exist throughout their life span. Such types of enzymes can be called as constitutive enzymes. These enzymes can be activated due to the attack of the xenobiotics. Plants need to act at a faster pace to detoxify these chemicals as they can get metabolized easily. It is important for us to understand the rate at which the detoxification occurs so that the selection of the herbicides can be done perfectly. The biochemical pathways are almost similar in the crops and weeds. In the case of crops, the metabolic rate is higher. The leaves will be covered with the waxy surface which can protect the plants from several chemicals. In order to enter the plant tissues, the herbicides should be lipophilic molecules. The oxidation of the metabolites results in the improvement of their polarity and reactivity. If any herbicide getsoxidized, the toxicity of the chemicals can be reduced.

The metabolism of the herbicides is based on the enzyme systems. There are four stages in which the herbicides act on the plant tissues. They are bioactivation, metabolic attack, conjugation and sequestration. The bioactivation stage generally occurs in the plant tissues. The herbicide which is in its inactive form can be called pro-herbicide. The detoxification of the herbicides can be done by the enzymes. These enzymes must have two primary functions. These enzymes must leave the herbicide inert. The enzymes must increase reactivity and polarity.

The pro-herbicides can be converted into phytotoxic compounds due to the bioactivation process. These pro-herbicides are inactive molecules and they get converted inside the plant cell. The metabolic cleavage of the bialaphos yields glufosinate. The metabolic attack can involve the addition of some of the chemical groups which are active. Groups such as carboxyl and hydroxyl groups are added. The plants disrupt the herbicides by the process of hydroxylation of alkyl groups and aromatic rings. The enzyme responsible for these reactions is cytochrome 450 oxidase. The P450 family contains several enzymes that are considered as a huge family in the entire plant kingdom. There is a porphyrin ring made up of the heam group. At the centerof this molecule,is an iron moiety. In metabolizing the herbicides, some of the conditions are to be met. The conditions are inhibition of p450 inhibitors, inhibition of carbonmonoxide, NADPH and oxygen requirement. Due to herbicide metabolism, these molecules will be linked with the metabolites. There will be a production of organic acids, sugars and amino acids. The processing of herbicide conjugates can start with glucosylation and malonic acid assisted sugar derivatization. Some of the malonylation reactions can be done by malaonyltranferases. Due to the malonylation process, the digestion due to glucosidases will be halted and this will help in the transport of conjugates.

Some of the chemicals can be added to the plants before the application of the herbicides. Because of these chemicals, the crops will develop tolerance towards the herbicides. These chemicals can be called as herbicide safeners. Some of these herbicide safeners can share the same structures. So there will be competition between the herbicides and the herbicide safeners for the site of action.

2.7 PHOTOSYNTHESIS

Plants can absorb light and convert it into chemical energy. The energy which is in excess can be used for the production of active oxygen species. The excess energy can also damage the membranes. This damage is termed as photodamage. In the photosystem II, D1 protein is present which is crucial in producing photodamage. There will be a damage repair cycle in the PS II reaction. The herbicides will act on the D1 protein. They can bind to the D1 protein. The crops and weeds can be influenced by the herbicides as the reactivity is the same in both plants. The diquat and paraquat are classified as bipyridinium compounds. They can be considered as herbicides and their mode of action is contact mode. Some of the useful properties of these compounds are that they can be water- soluble. They exhibit stability over wide pH ranges. They can be bound with the clay matter, organic matter and plants. They can accept one electron from photosystem I. They also accept number of electrons evolving from PS I. They can be oxidized by the molecular oxygen. They can be toxic to mammals, animals and plants.

The PS II herbicides can be absorbed by the roots. Some of the initial symptoms are chlorosis in the leaves. The chlorosis is predominant in the lower leaves. There will be damage to the root tips and margins. We can see both necrotic and chlorotic tissues once these herbicides are absorbed. Large concentrations of unsaturated fatty acids are seen in the thylakoids. Some of the unsaturated fatty acids are linolenic acids and linoleic acids.

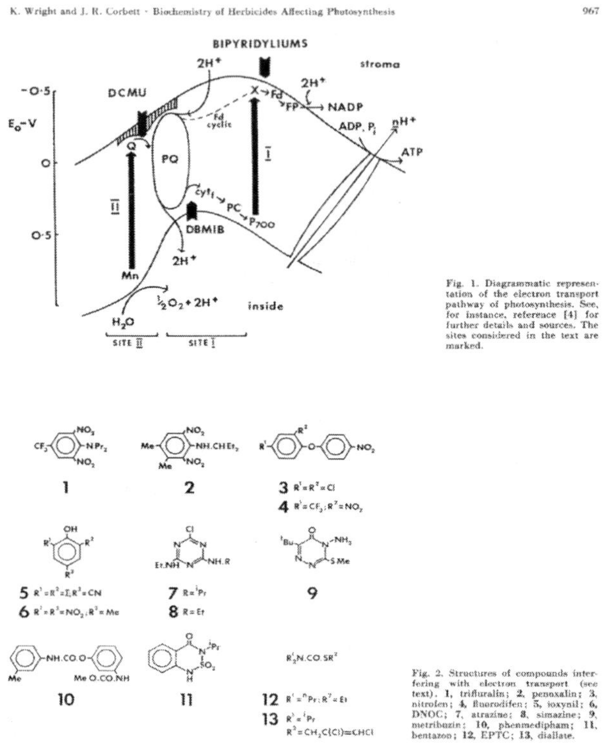

Figure 2.17. Effect of herbicides on photosynthesis

Source: https://www.degruyter.com/downloadpdf/j/znc.1979.34.issue-11/znc-1979-1118/znc-1979-1118.pdf

The effect of herbicides on the photosynthesis process is represented in figure 2.17. There will be the production of molecular oxygen due to photolysis. Active oxygen species can be considered important in determining the action of herbicides. The free radicals formed can have more unpaired electrons. The superoxide dismutase present in the chloroplasts can be considered as a metalloprotein. They will be present in the form of complexes with Fe, Mn and Cu-Zn. Hydrogen peroxide can inhibit the enzyme present in the stroma region. The survival of the plants can be based on the effectiveness of anti-oxidant protection and photo-oxidative stress.

Due to the deposition of excess excitation energy, the flow of electrons may not go beyond Q_A. This can result in many activated oxygen species and there will be excess fluorescence yield. Due to the attack of the free radicals, the tissues and cell membranes can be disintegrated.

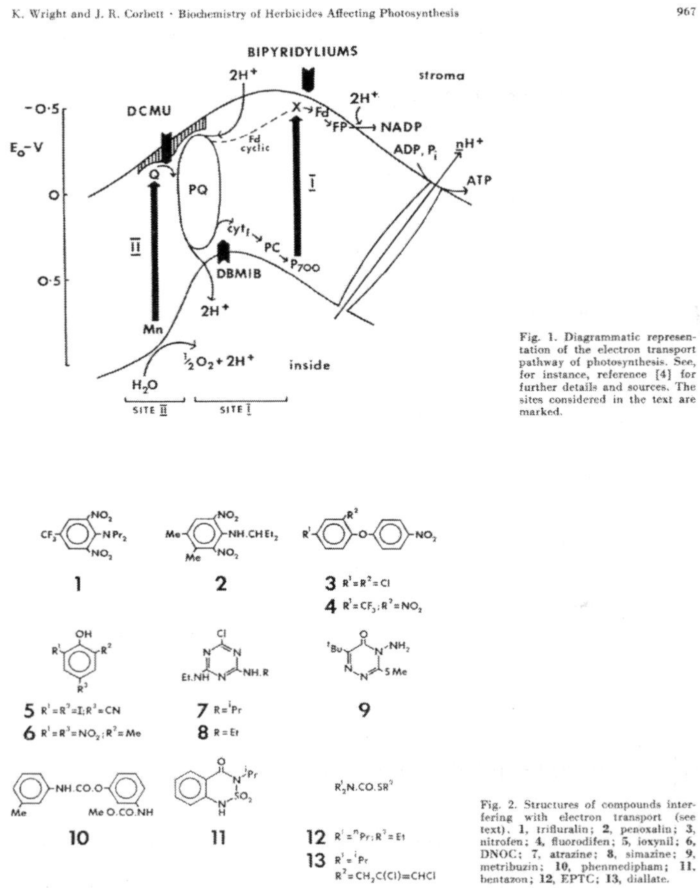

Figure 2.18. Molecules disrupting electron transport chain.

Source:https://www.degruyter.com/downloadpdf/j/znc.1979.34.issue-11/znc-1979-1118/znc-1979-1118.pdf

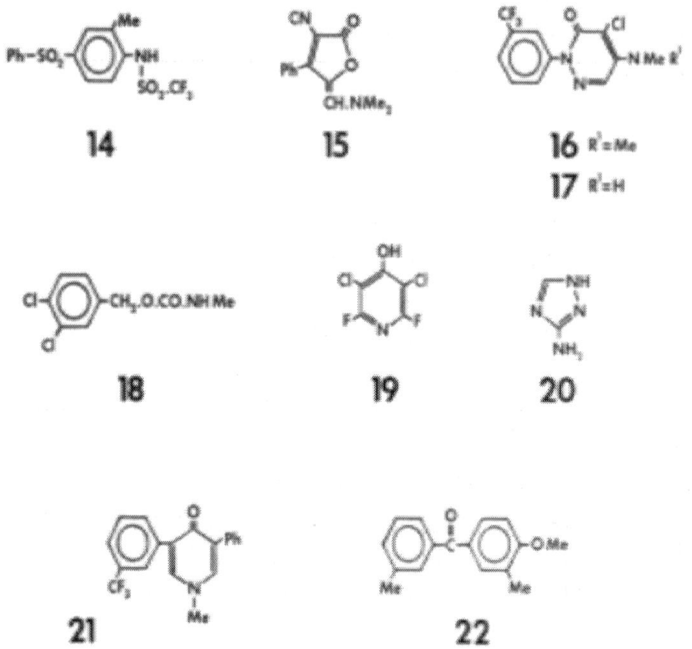

Figure 2.19. Molecules disrupting photophosphorylation.

Source:https://www.degruyter.com/downloadpdf/j/znc.1979.34.issue-11/znc-1979-1118/znc-1979-1118.pdf

The chemicals which influence photosynthesis are represented in figure 2.18. These chamicals represents: 1 - trifluralin, 2 - penoxalin, 3 - nitrofen, 4 - fluorodifen, 5 - ioxynil, 6 - DNOC, &- atarzine, 8 - simazine, 9 - metribuzin, 10 - phenmedipham, 11 - bentazon, 12 - EPTC and 13 - diallate. The molecules which act on the photophosphorylation process are represented in figure 2.19. These chamicals represents: 14 - perfluidone, 15 - difunone, 16 - san 6706, 17 - norflurazon, 18 - dichlomate, 19 - haloxydine, 20 - amino triazole, 21 - fluridone and 22 - methoxyphenone. The photosynthesis process is based on the light- harvesting nature of the carotenoid and chlorophyll pigments. These pigments can absorb solar energy in the thylakoids. The excited energy will be transferred into the reaction centers. Chlorophyll a can be considered as light- harvesting pigment. These molecules will have a head that absorbs light. This head will be on the magnesium ions. The tail of chlorophyll a molecules will get attached to the lipids and proteins. These pigments can be linked with the thylakoid membranes. Light can get absorbed by each chlorophyll molecule. One photon will be absorbed per second. The electron gets excited due to the absorption of a photon. Carotenoids and chlorophyll- b are present. The accessory pigments can absorb light of short wavelengths. The transfer of energy will be from the carotenoids to chlorophyll- b, chlorophyll- a and reaction center.

The herbicides can alter the biosynthesis of the carotenoids and chlorophylls. These herbicides can be termed as bleaching herbicides. There will be whitening of the leaves in the treated plants. In the process of photosynthesis, chlorophyll can be considered as an important element. In the reaction centers, these chlorophylls are present. This pigment plays an important role in the movement of electrons. The movement of electrons will be from water to $NADP^+$. The inhibitors of the chlorophyll biosynthesis can be called as peroxidizingherbicides. Diphenyl ethers can inhibit protoporphyrinogen IX oxidase. The peroxidizingherbicides can act at a faster pace. Their reaction will be reversible and they act as competitive inhibitors. They will be the same as there is protogenmolecule in their shape and other electronic properties. There will be the production of many oxygen species in the cytoplasm and stroma.

Protox can mediate the production of protoporphyrin IX molecule. This molecule can be called as photosensitizer. Some of the other molecules which have this activity are flavoproteins. These proteins can be resistant to degradation by proteases. In many cases, the carotenoids can safeguard the chlorophyll molecules from the active oxygen species. This can be done by dissipating the energy into heat through the quenching process. Some of the herbicides which can be crucial in controlling these weeds are phytoene desaturase inhibitors. Some of the examples of this type are diflufenican, flurochloridone and norflurazon. Fedtkeet al. in 2001 have described the action of diethylamines on lycopene cyclase. When plants were treated with the herbicides, the lycopene molecule can be stored in the plants. There will be reduced concentrations of beta carotene, lutein, violaxanthin and neoxanthin. The activity of these herbicides can be linked with the phytoene desaturase and lycopene cyclase.

Plastoquinone can be considered as an important electron acceptor. This molecule is crucial in the biosynthesis of carotenoids. This molecule is important in the electron transport mechanisms of photosynthesis. The beaching symptoms may appear if herbicides are sprayed. In general, Tyrosine amino acid which is aromatic via homogentisic acid can produce plastoquinone. The herbicides can lower the activity of HPPD or hydroxyl phenylpyruvate dioxygenase. The chemical isoxaflutole can be in the form of pro-herbicide. It is considered as a byproduct obtained due to metabolism. The herbicide diketonitrile can work as an effective inhibitor. Triketones or benzoyl-cyclohexanediones can be used as bleaching herbicides. The peroxidising herbicides can exhibit contact activity. Because of this activity, there will be poor selectivity. We can observe different metabolic rates in maize and weeds. The selectivity of the diphenyl ether was observed in rice and soybean crops.

2.8 AUXINS

2,4-D and phenoxyacetic acids were manufactured in the USA and UK. Several structural analogs were manufactured in the last three decades. Out of all the herbicides produced all over the world, 2,4-D has caught attention. It is a phenoxy herbicide. Some of the symptoms which appear on the application of these herbicides are that there will be increased auxin response and death will be imminent in most of the cases. The auxin type herbicides can be classified into 5 groups. They are benzoic acids, phenoxyalkanoic acids, pyridines and carboxylic acids with quinoline. The benzoic acids can include auxins which can form naturally. These auxins contain a free carboxyl group. There will be a negative charge due to the presence of the carboxyl group. Halogens can influence their activity and selectivity. 2,4,5-T can be considered as a powerful herbicide as it acts on the woody weeds which are tolerant to 2,4- D. Auxin-based herbicide can be effective in controlling weeds.

Indole-3-yl-acetic acid or IAA is one of the auxins which are important growth regulator. It helps in the division of the cells. Even the elongation and differentiation of the plant cells is based on IAA. Abscission, ripening, fruit setting, leaf senescence, apical dominance and another important process will be directly regulated with the auxins. It can be produced through tryptophan-independent and dependent pathways in the meristematic tissues. Auxins can stimulate and inhibit the growth of plants. The concentration of the auxins must be considered to evaluate their effectiveness. If the concentration of auxins is increased, there will be the production of ethylene. Because of this, the growth of plants will be inhibited. Ethylene production based on the auxin production can be crucial in defining the growth patterns of the plants. In the case of overproduction, we can expect inhibition of growth. Due to the production of ethylene, there will be enhanced production of abscisic acid or ABA. ABA is a plant hormone. Ethylene can produce senescence.

The closure of the stomata can be induced by ABA. The assimilation of carbon whilst the photosynthesis process can be hindered due to irregular closure of stomata. There will be formation of hydrogen peroxide due to photo-activation and hence we can expect a high rate of phytotoxicity in the weeds.

Auxins can get conjugated with myo-inositol, amino acids and glucose. During the germination of the seeds, the stored auxins can become activated and help in the growth process. Auxins need to bind with the receptor on the cells. The auxins bind with the receptors and make a strong bond within. Venis in 1985 explained that the binding process will be reversible. The binding process can be associated with the biological response. This process is restricted to the tissues which are very sensitive to the auxin molecules. The binding will be specific and saturable. The binding must be of higher affinity. A lot of research was done on TIR1 or transport inhibitor response 1 protein and ABP1 or auxin binding protein 1. In the etiolated maize coleoptiles, ABP1 auxins were observed in 1972 (reference). We can observe these molecules in the pteridophytes and bryophytes.

There are two important families which can regulate these herbicides. They are cyclohexanediones and aryloxyphenoxypropionates. These chemicals inhibit the ACCase enzymes or acetyl-CoA carboxylase. There are some other compounds that can control the long -chain fatty acids. Graminicides can be classified into 6 families. They are named as phenylpyrazolines, cyclohexanediones, aryloxypropionates, alaninopropionates, chloroacetamides and thiocarbamates. Dichlofop-methyl cannot interfere with nucleic acid synthesis or protein synthesis. This chemical can inhibit the production of lipids. The membranes present in the plants can contain fatty acids. They are important for both biochemical and structural roles. The fatty acids can be produced in the cytoplasm and stroma of the chloroplast. ACCase or acetyl CoA carboxylase is important in the production of fatty acids.

The chloroacetamides can be used to alter the biosynthesis of the fatty acids. This chemical can stop the elongation of the palmitate. The cuticle can be considered crucial in protecting plants from the pollutants and other chemicals. When there is a loss in the plasma membrane, the permeability can be disturbed. The hormone receptor functions and transport can be hindered. Once this occurs, the sites where the plasma membrane is disturbed can be the best sites for the herbicides. The chemical groups which can be active on the weeds and their lipid biosynthesis are oxiranes, oxyacetamides and chloroacetamides. The chemical chloracetamide is being used on rice, soybean and maize. The hydrolysis process can be important to determine the selectivity of the alaninopropionates. They can be used in the case of cereals.

The herbicides can target the sites where the biosynthesis of amino acids is prevalent. In 1981, the ammonium salt of the glufosinate was used to control the weeds (reference). The herbicides used in this process are categorized as post-emergence herbicides. This herbicide can inhibit glutamine synthase. Glufosinate can be produced naturally. The enzyme glutamine synthase can be used to convert glutamate to glutamine. The conversion can be mediated by ammonia and ATP. Glufosinate can inhibit the glutamine synthase in plants and bacteria. This herbicide can reduce photorespiration. This chemical can influence many pathways. They can alter the protein degradation process, influence peptide formation and membrane polarization are hindered. To control the weed populations, it is a good choice to use glyphosate which is also called as N-phosphonomethyl glycine. It is classified as a non-selective herbicide. We can classify this herbicide as post- emergence herbicide. Glyphosate can serve as a competitive inhibitor to PEP or phosphoenolpyruvate. It can bind to the site away from the active site of EPSP synthase. Since there is competitive inhibition, PEP may not bind to the site of its activity.

Glyphosate can result in necrosis, chlorosis and inhibits growth. Because of the inhibition, the plants will die. The symptoms will be hidden for atleast 10 days after its application. This chemical can increase phenylalanine ammonia lyase activity;there will be an inhibition of growth due to its accumulation in the plant tissues. Due to this chemical, there will be the production of necrotic areas in the leaves. This compound can be considered as non-selective herbicide. Some of the chemicals which have the same site of action are imidazolinones and triasolopyrimidines. Acetolactate synthase or ALS can be called as acetohydroxyacid synthase. This chemical can be used in the production of amino acids such as valine, isoleucine and leucine. These amino acids are branched. ALS inhibitors can exhibit herbicidal activities. They can regulate both perennial and annual grass. Total vegetation control is possible with imazapyr and sulfometuron. Bensulfuron-methyl can be used to protect rice crops. Chlorsulfuron can kill the weeds which attack cereals. ALS is localized in the chloroplast. It is nuclear- encoded. It is located at the site where the amino acid biosynthesis takes place.

3

Fungicides

CONTENTS

3.1 INTRODUCTION

Fungicides can be any biological agents or chemicals which can diminish the fungal growth. These agents can also kill the fungal spores. Mostly the contemporary fungicides can inhibit the fungal growth. Fungi can damage crops and so there will be higher economic losses. The fungicides can be useful in agriculture and they can even control the fungal infections in animals. Fungicides can also regulate oomycetes. Fungicides can be classified as contact fungicides, translaminar fungicides and systemic fungicides. The contact fungicides can influence the sprayed area and may not be absorbed by the plants. The translaminar fungicides can translocate after getting absorbed from the leaves. The translocation is possible with xylem and phloem bundles. If there is further growth of plants, the leaves can be protected. These fungicides will be available in the liquid form. These fungicides comprise of a huge concentration of sulfur. Some of the naturally available fungicides are jojoba oil, rosemary oil, and neem oil. In some cases, the bacterial species named Bacillus subtiliscan control the fungal spores and growth. Vinclozolin is a fungicide that is banned now for its ill effects on humans.

3.2 CLASSIFICATION

Based on their origin, fungicides can be divided into two groups. They were named as biofungicides and chemical fungicides. The biofungicides include microbes such as fungi and bacteria. This type of fungicides acts on the principle of antagonism. This type of action can be seen in the case of turf disease. Bacillus licheniformis and Trichoderma harzianum can be considered as biofungicides. Chemical fungicides are synthetic and produced from inorganic and organic chemicals. We can observe these chemical fungicides in the open markets. The chemical structure of the fungicides define their name. Some of them are triadimefon, fenariol, boscalid, flutolanil, hydrogen dioxide and so on. This naming technique was given by Burpee in 2006. These fungicides are divided into 16 groups. Some chemical fungicides are related to each other due to their chemical structures. Some of the chemicals which are closely linked are pyraclostrobin, trifloxistrobin and azoxystrobin. Strobilurin can regulate the growth of summer patch, gray leaf spot, brown patch and anthracnose. The chemical groups are triazoles, pyrimidines, strobilurins, polyoxins, benzimidazoles, dicarboxamides, phenylamides, carbamates, phosphonates, dithiocarbamates, aromatic hydrocarbons, peroxides, nitriles, phenylpyrolles, cyanoimidazole, carboxamides and biofungicides. Based on their topical activity, the fungicides can be classified into four groups. They are contact fungicides, localized penetrants, acropetal penetrants and systemic fungicides. The contact fungicides can be sprayed on the plant parts. They mainly act on the surface. These fungicides cannot be absorbed by roots, stems and leaves. If any plant is having an internal fungal infection, these fungicides may not control such a type of infection. Some of the examples of the contact fungicides are cyanoimidazoles, phenylpyrolles, peroxides, aromatic hydrocarbons, nitriles and dithiocarbamates. Localized penetrants are mainly absorbed by the leaves. They can be translocated in the plants through the vessels but the transport can be limited to less distance. The transportation of this fungicide between the leaves is not possible. These fungicides can regulate the fungal growth on the surface of the leaves and inside the leaf surface. Some of the localized penetrant fungicides are strobilurins and dicarboximides. The acropetal penetrants can enter the plant tissues via leaves, shoots and roots.

Common Name	FRAC[1]	Group Name	Trade Name(s)[2]	Topical mode of action[3]
azoxystrobin	11	Strobilurin	Heritage 50WG, Heritage TL, Heritage G	Acropetal Penetrant
boscalid	7	Anilene	Emerald 70EG (WDG)	Acropetal Penetrant
captan	M4	Thiophthalimide	Captan 80WP, Captec 4L	Contact
chloroneb	14	Aromatic Hydrocarbon	Terraneb SP, Teremec SP, Andersons Fungicide V, Proturf Fungicide V	Contact
chlorothalonil	M5	Nitrile	Andersons 5% Daconil 5G, ArmorTech 825 DF, ArmorTech CLT 720, ChloroStar 6F, Concorde SST 6F, Daconil 5G, Daconil Ultrex, Daconil WeatherStik, Daconil 2787, Daconil Zn, Echo 720F, Equus 500ZN, Manicure 6F, Manicure Ultrex, Lebanon Daconil 5G, Legend 6F, Pegasus L, Pentathalon	Contact
cyazofamid	21	Cyanoimidazole	Segway	Localized Penetrant
etridiazol (ethazole)	14	Aromatic Hydrocarbon	Koban 30WP, Terrazole 35WP	Contact
fenarimol	3	Demethylation Inhibitor	Rubigan AS	Acropetal penetrant
fludioxinil	12	Phenylpyrrole	Medallion 50WP	Contact
fluoxastrobin	11	Strobilurin	Disarm 480SC	Acropetal Penetrant
flutolanil	7	Carboximide	ProStar 70WP, Moncut 70-DF	Acropetal Penetrant
fosetyl-Al	33	Phosphonate	Autograph, Chipco Aliette, Chipco Signature, Lesco Prodigy Signature, Terra Aliette T&O	Systemic Penetrant
iprodione	2	Dicarboximide	Andersons Fungicide X, ArmorTech IP 233, Chipco 26 GT, Chipco 26019, Iprodione Pro, Lesco 18 Plus, Proturf Fungicide X, Raven	Localized Penetrant
mancozeb	M3	Carbamate	Dithane 37WF, Dithane T/O, Fore Flo-XL 4F, Fore Rainshield 4F, Formec 80W, Junction 61WF, Lesco Mancozeb 75DG, Lesco 4 Flowable Mancozeb 4F, Manzate 200, Pentathion, Protect DF, WingMan	Contact
maneb	M3	Carbamate	Maneb Plus Zinc F4, Maneb 75DF, Pentathlon 4F (75DG)	Contact
mefenoxam	4	Phenylamide	Andersons Pythium Control 1.2G, Mefanoxam, Quell, Ridomil, Subdue, Subdue MAXX	Acropetal Penetrant
metalaxyl	4	Phenylamide	Subdue2E, Subdue GR, Subdue WSP, Proturf Pythium Control, Apron (seed treatment), Verio	Acropetal Penetrant
metconazole	3	Demethylation Inhibitor	Tourney	Acropetal Penetrant
myclobutanil	3	Demethylation Inhibitor	Andersons Golden Eagle 0.39G, Eagle 20EW, Siskin	Acropetal Penetrant
PCNB	14	Aromatic Hydrocarbon	Andersons FFII 15.4G, Defend 4F, Engage 75W, Fluid Fungicide II, Lesco Revere 4000 4F (10G), Parflo 4F, PCNB 12.5G, Penstar 75WP, Terraclor 400F (75WP), Turfcide 400F (10G)	Contact
phosphites (salts)	33	Phosphonate	Alude 5.2F, Biophos, Fosphite, Magellan 6.69F, Phostrol, Reliant, ReSyst 5F, Vital 4L	Systemic Penetrant
polyoxin D zinc	19	Polyoxin	Endorse 2.5WP, Affirm 11.3WDG	Localized Penetrant
propamocarb HCl	28	Carbamate	Banol 6S	Localized Penetrant
propiconazole	3	Demethylation Inhibitor	ArmorTech PPZ 143, Banner GL 3.6WSP, Banner MAXX, Kestrel, Lesco Spectator 3.6EC Propiconazole Pro 1.3MEC, Strider	Acropetal Penetrant
pyraclostrobin	11	Strobilurin	Insignia 20WDG	Localized Penetrant
tebuconazole	3	Demethylation Inhibitor	Torque	Acropetal Penetrant
thiophanate-methyl	1	Benzimidazole	Absorb TM, Allban, Andersons Systemic Fungicide 2.3G, ArmorTech TM 462, Cavalier 2G (4.5F, 50WSB), Cleary's 3336 (2G, 4F, 50WSB, 70EG), Fungo Flo 4.5F, Fungo 50WSB, Lesco T-Storm 50WSB, Proturf Systemic Fungicide, Systec 1998 85WDG, T-Bird WDG, Tee-Off 4.5F	Acropetal Penetrant
thiram	M3	Carbamate	Defiant 75WDG, Lesco Thiram 75DG, Spotrete 4F	Contact
triadimefon	3	Demethylation Inhibitor	Accost 1G, Bayleton 50WSP, Andersons Fungicide VII 0.59G, Andersons 1% Bayleton 1G Lebanon Bayleton 1G, Lesco Systemic Fungicide 50WDG	Acropetal Penetrant
trifloxystrobin	11	Strobilurin	Compass 50WDG	Localized Penetrant
triticonazole	3	Demethylation Inhibitor	Trinity, Triton	Acropetal Penetrant
vinclozolin	2	Dicarboximide	Curalan 4F (50EG, WSB, 50DF), Lesco Touche EG, Vorlan 50DF	Localized Penetrant

Figure 3.1. Different types of fungicides.

Source: https://www.doyourownpestcontrol.com/SPEC/technical-bulletin/UMassfungicide_class_mode.pdf

Figure 3.1 explains various types of fungicides.This type of fungicides can be translocated through the xylem. They can lower the growth of fungi, especially on the treated surfaces. They can exhibit their activity in the vicinity of the treated surfaces also. Some of the examples of these fungicides are fluoxastrobin, azoxystrobin, acylalanines, carboxyamides, triazoles and benzimidazoles. The fungicides which can be absorbed into the vessels such as xylem and phloem so that they can get translocated both upwards and downwards in the plants can be called systemic fungicides. They can reduce fungal growth. Phosphonates can be considered as systemic fungicides. When seen under a microscope, the thallus of the fungi can be appearing like hyphae. The fungicides can be classified into two groups. They are classified into site- specific inhibitors and multi-site inhibitors. The site- specific inhibitors can work on specific sites in the fungal cell. The multisite inhibitors can be active on several sites.

Pre-mixed Products (Common Name)	FRAC[1]	Trade Name(s)[2]
azoxystrobin + propiconazole	11 plus 3	Headway
chlorothalonil + propamocarb	M5 plus 28	Banol C, Lesco Par,
chlorothalonil + propiconazole	M5 plus 3	Concert, Echo Propiconazole Turf Fungicide
chlorothalonil + thiophanate-methyl	M5 plus 1	Broadside, ConSyst, Spectro, Peregrine
chlorothalonil + fludioxonil + propiconazole	M5 plus 12 plus 3	Instrata
chlorothalonil + triticonazole + stress guard	M5 plus 3 plus ?	Reserve
fluoxastrobin + chlorothalonil	11 + M5	Disarm C
fluoxastrobin + myclobutanil	11 + 3	Disarm M
mancozeb + copper hydroxide	M3 plus M1	Junction
mancozeb + myclobutanil	M3 plus 3	MANhandle
propamocarb + fluopicolide	28 plus 43	Stellar
pyraclostrobin + boscalid	11 plus 7	Honor
thiophanate-methyl + chloroneb	1 plus 14	Proturf Fungicide IX
thiophanate-methyl + flutalonil	1 plus12	SysStar
thiophanate-methyl + iprodione	1 plus 2	ArmorTech TMI 20/20, 26/36 Fungicide, Dovetail, Lesco Twosome, Proturf Fluid Fungicide
thiophanate-methyl + mancozeb	1 plus M3	Duosan WP, Duson WSB
triadimefon + flutolanil	3 plus 7	Prostar Plus
triadimefon + metalaxyl	3 plus 4	Proturf Fluid Fungicide II
triadimefon + thiram	3 plus M3	Proturf Fluid Fungicide III
triadimefon + trifloxystrobin	3 plus 11	Armada
triadimefon + trifloxystrobin + stress guard	3 plus 11 plus ?	Tartan

Figure 3.2. Pre-mixed fungicides.

Source: https://www.doyourownpestcontrol.com/SPEC/technical-bulletin/UMassfungicide_class_mode.pdf

Figure 3.2 represents the pre-mixed fungicides along with the trade names.

3.3 TOXICOLOGY

The fungicides will show certain characteristics. They exhibit lower toxicity in mammals. They show a low rate of ecotoxicity. Phytotoxicity due to fungicides is low. In the case of mycelia and spores, these fungicides can penetrate easily. There will be less biodegradation of these fungicides on the surface of the leaves. Triazoles are being used in the market since the 1980s. Some of the examples of triazoles are triticonazole, triadimefon, tetraconazole, tebuconazole, propiconazole, myclobutanil, febuconazole and difenoconazole. These fungicides can act on many diseases that infect ornamental plants, field crops, vegetables and turfgrasses. We can use these fungicides as sprays and in the treatment of seeds. We can use them as agents for curative treatment. The application of triazoles should start before the fungal infection. The triazoles can translocate through the xylem. They can be absorbed by the leaves. Mostly these fungicides can migrate in the leaf tissues. These fungicides can disrupt the sterol biosynthesis. This component is important for the stability of the cell membranes of fungal members. Triazoles can exhibit lower toxicity. There will be some rate of irritation in the throat, lungs and nose. Cyproconazole can cause less damage to the bees, mammals and birds. They can exhibit minimum toxicity. The aromatic hydrocarbon fungicides can be considered as crucial in the management of fungal infections. Some of the examples of this fungicide are sodium pentachlorophenate, pentachlorophenol, quintozene, hexachlorobenzene, chloroneb and chlorothalonil. Some of the problems with this fungicide is that it causes liver damage, heat production and eye problems. Chloroneb can be considered as a broad spectrum fungicide. It can be classified as a systemic fungicide. Due to this fungicide, there will be limited DNA polymerization. We can use this fungicide in treatment of turfgrass to control pythium and typhula. This chemical can harm fishes and even animals. Ziram, zineb, thiram,nabam, ferbam and mancozeb can be classified as dithiocarbamate fungicides. Thiram can be used to regulate or even control crop damage.

Generic Names	Chemical Group	
propiconazole triadimefon myclobutanil triticonazole tetraconazole	triazoles	DMI (demethylation Inhibitors fungicides)
fenarimol	pyrimidines	
fluoxastrobin trifloxystrobin azoxystrobin pyraclostrobin	strobilurins	
polyoxin D	polyoxins	
thiophanate-methyl	benzimidazoles	
iprodione vinclozolin	dicarboxamides	
mefenoxam	phenylamides	
propamocarb	carbamates	
fosetyl aluminum phosphonate	phosphonates	
mancozeb thiram	dithiocarbamates	
quintozene chloroneb ethazole	aromatic hydrocarbons	
hydrogen dioxide	peroxides	
chlorothalonil	nitriles	
fludioxonil	phenylpyrolles	
cyanofamid	cyanoimidazole	
flutolanil boscalid	carboxamides	
Ecoguard Sonata Soilguard	Biofungicides	

Figure 3.3. Chemical groups.

Source: https://www.researchgate.net/publication/221905691_Introduction_and_Toxicology_of_Fungicides

The fungicides are classified into chemical groups and they are represented in figure 3.3.

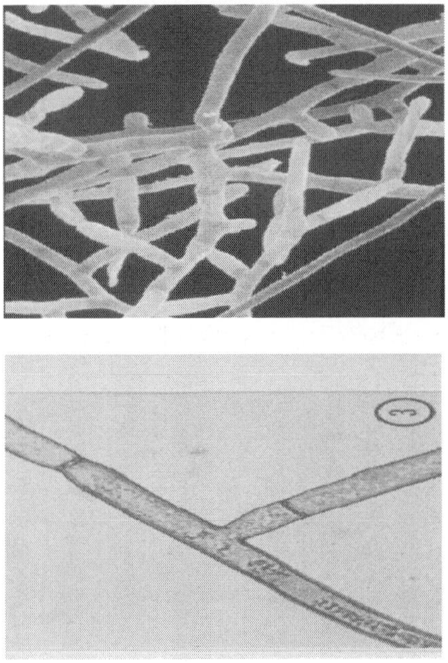

Figure 3.4. Fungal hyphae.

*Source:*https://www.researchgate.net/publication/221905691_Introduction_and_Toxicology_of_Fungicides

The fungicides will act on the fungal hyphae. They are shown in figure 3.4.

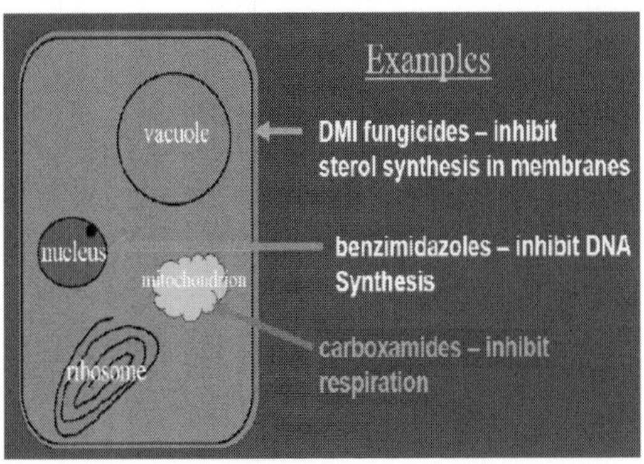

Figure 3.5. Site specific inhibitors.

Source: https://www.researchgate.net/publication/221905691_Introduction_and_Toxicology_of_Fungicides

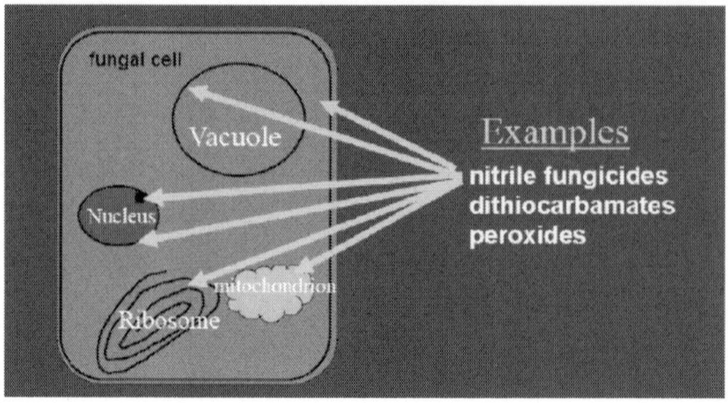

Figure 3.6. Mult-site inhibitors.

Source: https://www.researchgate.net/publication/221905691_Introduction_and_Toxicology_of_Fungicides

The fungicides can act as site-specific inhibitors and multi-site inhibitors. (Figure 3.5& 3.6)

Common name	Rat oral LD 50	Rabbit dermal LD50
Difenoconazole	1,453	2,010
Fenbuconazole	>2,000	>5,000 (rat)
Mycobutanil	1,600	>5,000 (rat)
Propiconazole	1,517	>4,000
Tebuconazole	1,700	>2,000
Triadimefon	569	2,000
Triticonazole	>2,000	>2,000

Figure 3.7. Mammalian toxicity observed while using triazole fungicides.

Source: https://www.researchgate.net/publication/221905691_Introduction_and_Toxicology_of_Fungicides

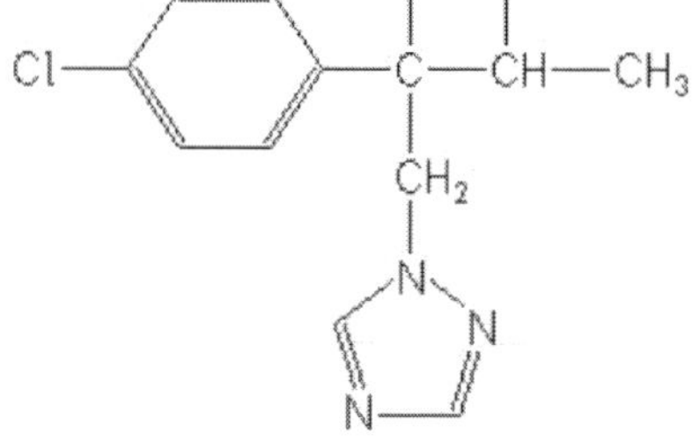

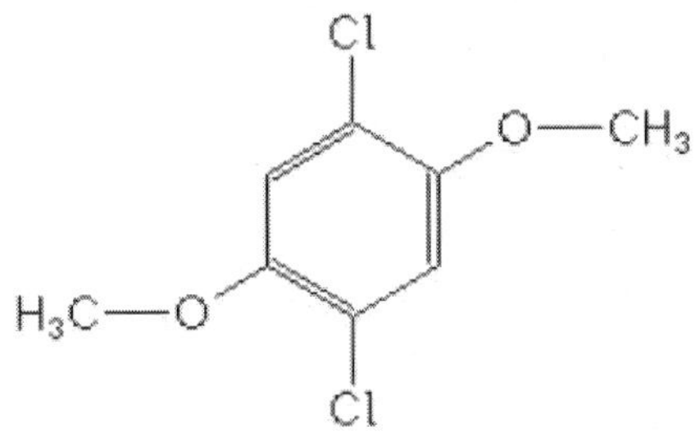

Figure 3.8. Cyproconazole (Top) and Chloroneb (bottom).

Source: https://www.researchgate.net/publication/221905691_Introduction_and_Toxicology_of_Fungicides

The chemical structures of the fungicides cyproconazole and chloroneb are shown in figure 3.8.

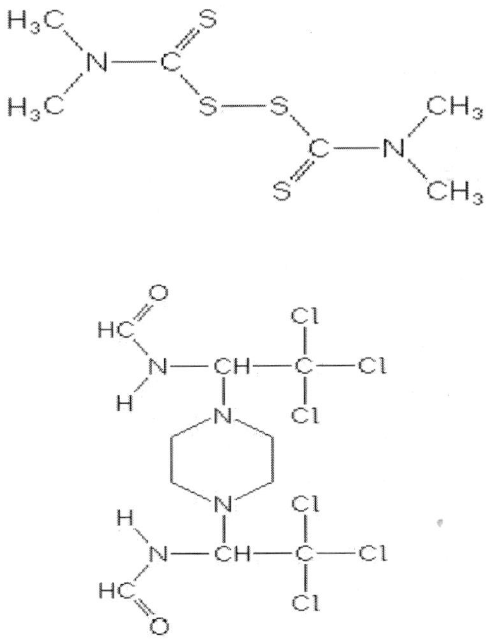

Figure 3.9. Thriam(Top) and Triforane (bottom).

Source:https://www.researchgate.net/publication/221905691_Introduction_and_Toxicology_of_Fungicides

The chemical structures of thiram and triforaneare shown in figure 3.9.

3.4 RESTRICTIONS AND LABELING

The fungicides must be labeled so that the worker can be protected from the exposure. Nobody should enter the field atleast 24 hours post spraying. It is mandatory to wear protective clothing to avoid any exposure from the fungicides. We need to label the fungicides so that the restrictions can be followed easily. One should read the label properly so that he or she may not spray unnecessary chemicals to get the crop damaged. Triphenyltin hydroxide can be considered as a restricted fungicide. So it should be labeled as restricted use pesticide. Most of these fungicides can be available in the form of solids. They cannot be soluble in water in normal conditions. It is crucial to prepare formulations. Some fungicides can be used as wettable powders and they should be labeled as WP. They comprise of wetting agent and solid fungicide. If we add water to this wetting agent, it becomes a suspension. These suspensions can easily settle. It is important to use agitator so that the spraying can be easier.

There are some fungicides that can be available in the open market as water -soluble pouch containers. These pouches can be dissolved in the sprayer tank and so the fungicide will be released. There is no need for any mixer or agitator. Powdering is not required in this case. We can label this type as WSP. Some fungicides can be defined as powders or D. They are powders that can be combined with ingredients that can be inert or non-reactive. The active materials, in this case, are kept at a minimum. We can use powder type fungicides in the home garden and other applications. If the fungicides are incorporated into granules, then we call such as type of fungicides as granules. The materials which can be used as granules in this case can be clay. These granules can be merged with the soils. Some fungicides cannot be dissolved in water and it is crucial to know that these fungicides can be dissolved in organic solvents. When these fungicides are mixed with water, there will be the formation of an emulsion. It is nothing but solvent drops suspended in water. We can expect a milky appearance.

Some fungicides can be subjected to wet grinding process. The fungicides can be insoluble and they should be grounded into fine particles. These types of fungicides can be called as flowables. These fungicide particles can be colloidal in nature and size. Once they get suspended in water, they appear as a thick liquid. We can use these flowables for long periods. It is essential to mix these fungicides well before use. There are some advantages to the use of flowables. They are easy to mix and mostly powder free. They exist as suspensions for more time than the wettable powders. The dry flowable formulations can be called as dispersible granules. Mostly they exist as minute granules. They can be stored in normal conditions and freezing is not required. These fungicides can be used in the form of suspensions. They may not be contaminated with powder or create powder. Penncozeb DF and Manzate 75 DF can be considered as examples for dispersible granules. They can be labeled as DG. If any liquid fungicides turns into gas, they are definedas fumigant. They can be used in fumigating soils.

Most of the fungicides can act on the exterior surface of the host. This process is called protection. If any fungicide acts on the outer surface, we call it as protectant fungicide. Most of the fungicides can be sprayed on the leaves and the fruits. The chemical action offered by the fungicide on the host can be called as therapy. The therapy action can be done inside the host tissues. Triadimefon or bayleton can be used to control the leaf rust in wheat crops. This chemical can also protect the plants from the fungal spores. These spores can grow and produce disease to the wheat crops. The fungicide named propiconazole can be active against several fungal diseases and can stay active for several days after its application. These protectant fungicides can be stable mostly but there are some exceptions. They cannot dissolve in water easily. They can be more toxic to the fungal members. The addition of water to this chemical cannot create any change in its chemical structure or removes active species. The fungicides are capable of producing many toxic reactions that can kill the fungal members. Some fungicides can diminish the respiration process. Some can damage the nuclear division machinery of the spores. They can fuse with some metals which are needed for the regular activity of the fungal members. The metal ions will not be available to the metabolic activities of the fungal species. The cell permeability of the fungal species can be destroyed due to these fungicides. Due to this activity, the fungal cells cannot hold the nutrients for longer periods. Some of the fungicides can alter the formation of the cell wall. Triadimefon and propiconazole are the fungicides that can have this activity on the rusts.

Some of the fungicides can produce a high rate of toxicity so that even other species can get damaged or killed. In the case of unfavorable environmental conditions, the fungus will get killed due to fungicides effectively. This process can be defined as phytotoxicity. The toxicity of the pesticides can be classified as oral toxicity and dermal toxicity. Oral toxicity occurs if the pesticides are absorbed through the mouth. If the pesticides are absorbed via the skin, we can call it as dermal toxicity. The dermal toxicity can be treated with the utmost care. The chemicals which act as fungicides must be handled carefully so that there should not be any spilling and splashing. It is important to wear a mask to avoid unnecessary exposure. When these chemicals join with others, there will be the formation of compounds that can produce higher toxicity. The process of increase in toxicity of the fungicides can be called as potentiation. It is important that we need to ensure the complete rinsing of the containers. The chemicals must not be added to the spray tanks. It is a good practice to keep a record of all the chemicals which are being used. The chemicals which are to be handled must also be labeled and recorded.

There are some chemicals that can harm humans. There are some symptoms associated with the harmful effects of these fungicides. Each pesticide or fungicide can be differentiated with the symptoms which they produce. Some of the symptoms which are linked with these pesticides are dizziness, vomiting, sweating, nervousness, etc. The fungicides should be labeled according to their chemical name. Some fungicides can be made with some formulations. There are diverse applications with these pesticides. Because of the nature of the chemical compositions, these chemicals can develop certain incompatibility. The pesticides or fungicides can vary in their mobility and persistence. The pesticides which are defined as mobile and persistent can contaminate groundwater. In the case of coarse -textured soils, we can expect minimal organic matter and so the risk of groundwater contamination with pesticides is common. The fungicides can be classified as immobile in the case of clay soils as they can adhere to the clay particles. There will be a generation of organic matter in the clay soils which can render it impregnable.

In the case of cereals, the seeds can be treated with fungicides. They can safeguard the seeds from the rotting process. They can also protect seeds from blight. The fungicides will protect the seedlings from attack by the pathogens present in the soils. Due to unfavorable conditions, seedling can suffer from lack of resources and so the growth will be limited which can create vulnerability. This can be subjected to fungicidal treatment. Protectant fungicide can be used to treat seeds from the pathogens. Many of these fungicides can protect the seeds from the blight disease. The seeds can be treated with the fungicides and they offer protection from other diseases too.

Some of the diseases which can be controlled by the fungicides are loose smut, scab, semi-loose smuts and black point. The loose smut of oats can be successfully treated with fungicides. Most of the barley varieties can be prone to fungal attacks. Chickpeas can be susceptible to the infection caused by pythium. Treating chickpeas with the fungicides can result in a protection against ascochyta. Treating soybeans and dry beans with fungicides can lower the blight disease in case unfavorable conditions prevail. Flax seeds can be treated with the fungicides so that protection against seedling blight, damping off and seed rot is possible. The flax seeds may be treated with the fungicides to regulate pasmo infection.

In the case of potatoes, the fungicides can be effective against Fusarium, Rhizoctonia and Pythium. Most of the fungicides cannot offer protection against clavibacter and erwinia. Safflower rust is the main problem with safflower. The main diseases infections seen in safflower are footrot, root rot and blight. The safflower rust is caused due to infection progress in the seeds. The mildew infections are mostly soil-borne. Some of the fungicides used in the treatment of seeds are mefenoxam and metalaxyl. These fungicides are no more effective. The seeds can be treated with these fungicides. The seed of the forage legume can be treated with these fungicides. The foliar fungicides can be used to regulate fungal diseases. These chemicals can protect the crop quality and yield. Thiophanate methyl and thiabendazole can be classified as systemic fungicides. They are classified under benzimidazole fungicides. Because of the frequent use of these fungicides, there is a chance that the pests develop resistance.

The spraying process can be performed in many ways. We can use both air equipment and ground equipment. The size of the droplet can range between 200 to 400 microns. The application of these fungicides should be less than 6 feet from the crops. In the case of plants that have a waxy surface and hairy surface, these fungicides can be deposited as droplets. We can explain this phenomenon in the case of cabbage and wheat. The fungi can be resistant to certain fungicides. Cercosporawhich cause leafspot fungus in sugarbeet can develop resistance to thiophanate methyl, thiabendazole and benomyl compounds. These chemicals are classified as benzimidazole fungicides. Fusarium sambucinum and Helminthosporiumsolani have developed resistance to fungal populations. Biofungicidesare shown in figure 3.10

Product	Active Ingredient	Disease	Treatment Site
Ballad	*Bacillus pumilus*	Several (Foliar)	Foliar
Bio-Save	*Pseudomonas syringe*	Post-harvest	Irish and sweet potatoes in storage
Contans	*Coniothyrium minitans*	White Mold	Soil applied
Kodiak	*Bacillus subtilis*	*Pythium, Rhizoctonia, Fusarium*	Seed treatment, beans only
Mycostop	*Streptomyces griseoviridis*	Several	Greenhouse; Soil applied
Regalia	Plant extract	Powdery mildew	Foliar
RootShield Granules, RootShield WP	*Trichoderma harzianum*	*Pythium, Rhizoctonia, Fusarium*	Soil applied
Serenade	*Bacillus subtilis*	Powdery mildew, other foliar diseases	Foliar
T22-HC	*Trichoderma harzianum*	*Pythium, Rhizoctonia, Fusarium*	Soil applied
Surround	Kaolin	Powdery mildew	Foliar
Trilogy	Neem Oil	Powdery mildew	Foliar
Actinovate AG	*Streptomyces lydicus*	*Fusarium, Rhizoctonia, Pythium, Phytophthora*	Soil applied Foliar
SoilGard	*Gliocladium virens*	Damping off	Greenhouse-transplants, soil applied

Figure 3.10 Biofungicides

*Source:*https://www.researchgate.net/publication/221905691_Introduction_and_Toxicology_of_Fungicides

3.5 EFFECTS OF FUNGICIDES

Oomycetes and fungi can be classified as agents which can produce plant diseases. The fungal growth can be regulated by the fungicides, but complete removal of fungal growth is not possible.

Fungicides, Formules	Structures	Primary mode of action in fungus
Copper and/or sulfate		
Bordeaux mixture $CuSO_4 \cdot 3Cu(OH)_2$ $3CaSO_4$		Inhibit several enzymes involved in respiration and this multi-site action (Leroux 2003)
Kumulus S_x		Inhibit several enzymes involved in respiration and this multi-site action (Leroux 2003)
Cuproxat $3Cu(OH)2CuSO_4$		Inhibit several enzymes involved in respiration and this multi-site action (Leroux 2003)
Strobilurine		
Kresoxim-methyl $C_{18}H_{19}NO_4$	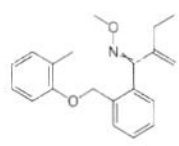	Inhibits spore germination and mycelial growth (Leroux et al. 2002) QoI inhibitors, which act to inhibit the respiratory chain at the level of Complex III (Grossmann et al. 1999)
Azoxystrobin $C_{22}H_{17}N_3O_5$	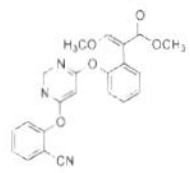	Inhibits spore germination and mycelial growth (Zhang et al. 2010)
Pyrrole		
Fludioxonil $C_{12}H_6F_2N_2O_2$	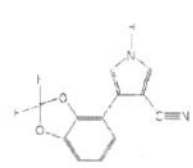	Perturbe the osmo-regulatory signal transmission pathway (Pillonel and Meyer 1997) Inhibits spore germination and mycelial growth (Leroux et al. 2002)
Anilides		
Fenhexamid $C_{19}H_{15}FN_2O_4$	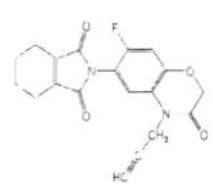	Inhibits germ-tube elongation and mycelial growth (Hänβler and Pontzen 1999) A sterol biosynthesis inhibitor: inhibits the 3-keto reductase involved in the enzymatic complex of the sterol C-4 demethylation (Debieu et al. 2001)
Phthalimide		
Folpet $C_9H_4Cl_3NO_2S$	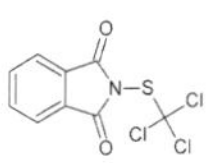	Inhibits spore germination and mycelial growth Multi-site action (Teisseire et al. 1999)
Azoles		
Cyazofamid $C_{13}H_{13}ClN_4O_2S$	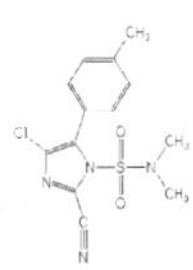	Inhibit mycelial growth Indirect effect against germination of zoosporangia, zoospore motility, cystospore germination, and oospore formation Impairment of the ATP energy generation system Inhibit several enzymes involved in respiration (Mitani et al. 2001)
Paclobutrazol $C_{15}H_{20}ClN_3O$	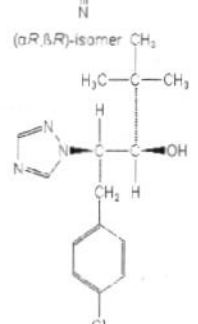	Inhibition of sterol biosynthesis at the 14α-demethylation step (Haughan et al. 1988)
Triadimefon $C_{14}H_{16}ClN_3O_2$	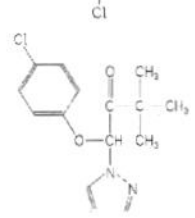	Sterol C-14 demethylation inhibitor (Debieu et al. 1995)

Figure 3.11. Chemical characteristics of fungicides

Source: https://www.researchgate.net/publication/221803317_Fungicide_impacts_on_photosynthesis_in_crop_plants/link/0fcfd50c061d089138000000/download

This property can be called a eradicative effect. The application of fungicides to regulate the fungal populations after infection can be called a curative effect. In the case of curative action, the fungicides should be made contact with the plant body. The triazole fungicides can be used to cleave the intercellular mycelia in the rose leaves. The crown rots and root rots cannot be controlled with the fungicides. The crown and root rots can be effectively treated if these fungicides are subjective to preventive chemical control. We can efficiently control the growth of rhizoctonia and phytophthora. We can use chemicals to control the wood rots, cankers and vascular wilts. Curative treatments can be used to regulate Fusarium species. One of the most important species which can be studied is Fusarium oxysporum. Carbendazim and prochloraz can be used to regulate the fusarium wilt.

3.6 PLANT PATHOGENS

The fungicides can diminish the infection rates in the plants. The pathogens can be classified as obligate parasites and non-obligate parasites. The non-obligative parasites can be called as facultative pathogens. The obligate parasites can reproduce and even grow in large numbers in the hosts. The facultative pathogens can survive on the dead materials. We can even use microbial artificial media to grow this type of pathogen. The pathogenic species include fungi, bacteria, algae, viruses, nematodes, viroids, phytoplasmas and trypanosomatids. We can also consider the species like plasmodiophora and oomycota as pathogenic organisms.

Fungal species can be considered as important groups which can inflict diseases in plants. Some of these groups can be called supergroups. The supergroups are archaeplastida, opisthokonta, rhizaria and chromalveolata. Most of the fungal species can be considered as heterotrophic in nature. They can ingest nutrients which are having low molecular weights. These pathogens can appear as filamentous species. They can grow hyphae with a tip. These hyphae can bedifferentiated by the septae. Yeast is a single- celled organism that can divide through cell division. These organisms will have cell walls that are rigid. Oomycetes on other hand,have cell walls based on cellulose. Sterol ergosterol is often found in the cell walls of fungi. Most of the fungal members will grow with the help of spores. They have both sexual and asexual structures. The asexual reproduction will be through spores. The fungal species which cannot produce spores can be called 'Fungi imperfecti'. We can call it as deuteromycota. These are linked with the fungal members. The fungal species are classified into 6 main groups. Out of these 6 groups, Ascomycota and Basidiomycota can be considered important. Chytridiomycota is the fungal group that harbors most of the pathogens. These species can be called as Chytrids. They can infect animals and in some cases human beings. The lucerne pests and maize pests can be considered examples of chytrids. Ascomycota is considered as a biggest phylum. Most of the pathogens belong to this phylum. In yeast group, there are some species which can cause diseases. Yeasts can be used in the iindustrieswhich work on fermentation processes. Necrotrophic effectors can be produced by the class pleosporales. This class belongs to Ascomycota.

The basidiomycota members can be classified into 2 groups. They are pucciniomycotina and ustilaginomycotina. These groups have created much damage to the plant communities. The smuts and bunts belong to the Ustilagogenus. These pathogens can be considered as biotrophic organisms. Most of the rust diseases are caused by pucciniomycotina. These rusts can be considered as biotrophic pathogens. The main problem with these organisms is that they cannot be grown on artificial media.Most of the fungal pathogens belong to the oomycota group. The late blight of potato can be considered as destructive diseases caused by Phytophthora infestans. This has caused Irish famine. Some of the other groups which are considered important are peronosporales and pythiales. The downy mildews are caused by peronospora. Seedling damping-off is caused by pythium. The fungal species are heterotrophic and eukaryotic. Clubroot is a root disease caused by plasmodiophora. The pathogens were classified into biotrophs and necrotrophs. If a pathogen acquires nutrients from the host which is living, then we can define such a pathogen as biotroph. If pathogens survive on dead matter, then we can call it as necrotroph. The biotrophs can be considered as host- specific. The resistance is caused due to the storage of jasmonic acid.

3.7 RESISTANCE

The resistance against the fungicides can be important in agriculture. Bioassay and field assay are crucial in determining the resistance of plants towards fungicides. Cuomo et al.in 2007 have explained that the genomes present in the fungal members are plastic. They comprise of several polymorphisms. The fungicides which can act on several biochemical steps can be called as multisite inhibitors. The mutations must occur so that the resistance process can occur and often this process is slow. The properties of pathogens can influence resistance and evolution. The resistance is based on the biochemical, epidemiological and genetic factors. The biochemical factors will be based on the mechanisms required for resistance. They are also based on some of the biochemical steps which can be disruptible. The epidemiological properties involve or depend on the isolation of the pathogen populations. Because of this, the species which will create competition will be stopped at the entry points. This is based on the strength of infection and it will be on all the crop stages. It is based on the life cycle of the pathogen. Some pathogens can live for short periods and some for longer periods. The sporulation of the pathogens can be crucial in determining the epidemiological properties. The method of dispersion can be crucial in determining the resistance offered by the plants. If we consider genetic properties, the plants which offer resistance must have resistance alleles. The asexual and sexual reproduction may affect the inheritance of the resistance properties. The genotypes should be fit. The genotypes should be relatively abundant.

The fungicides can influence resistance through their biochemical activity and physicochemical activity. The mode of application can be important. The biochemical property involves the interaction with the target which can be susceptible. The toxicological or the physicochemical activity can be based on the transport and partition properties. It is based on the polarity, volatility, solubility and stability of the fungicides. The mode of application of the fungicides is also important. We can integrate the fungicides with other disease management tools. The disease management tools involve biofungicides, crop hygiene and crop rotation. The area which is to be treated can be crucial in defining the resistance to the fungicides. The formulation and dose in which the fungicides are sprayed can be crucial.

The resistance can be classified into innate and acquired resistance. The acquired resistance can be based on the fungal pathogens. The resistance offered by the plants against these fungicides has becomes the greatest problem among the growers. Fungi can adapt easily with a number of pesticides acting on it. The resistance can be due to the activity of minor genes which we call it polygenic resistance. If the mutations in the major genes bring about resistance, we call it as monogenic resistance. The monogenic resistance can be called as oligogenic resistance. The polygenic resistance can be induced by the cytoplasmic genes and nuclear genes. The resistance offering genes can be expressed immediately in the haploid fungi and inthe case of diploid fungi, the resistance genes may not be expressed soon as there will be both dominant and recessive genes. The allelic variants can produce diverse levels of resistance. It is important to consider the heterokaryosis as this process can produce resistance in fungal members against fungicides. Some of the novel tools such as site-directed mutagenesis, sequencing and gene cloning can offer resistance in the fugal members.

Acquired resistance can occur because of the genetic modifications which can be inheritable. Because of this resistance, the fungicides which are active and toxic on particular or specific fungi may not be active anymore. Some of the fungicides which can be used to regulate fungal growth are dithiocarbamates, copper derivatives and sulfur compounds. These fungicides can act on multiple sites and so there will be multiple target sites. Due to this activity, they are important in controlling pest fungi. Some of the important factors which are to be considered in explaining the fungicidal resistance are as follows. The relationship between dominant and recessive among the wild types and resistant type must be established. The allelic variants present in a locus must be enumerated so that we can identify if they are involved.

Crucial factors which form the basis of the resistance can be found in the chromosomes. The genes which confer resistance can be placed or identified in the extra chromosomal pieces. We can demarcate the cytoplasmic genes and nuclear genes with the help of inheritance patterns. In case of genes from the cytoplasm, we can expect non-Mendelian inheritance. This type of inheritance can be designated with the uniparental transmission. In the case of nuclear genes, the zygotes if formed contain each allele from both the parents. So it may be follow the Mendelian inheritance. The genes responsible for conferring resistance are identified in the nuclear chromosomes. The resistance in the fungal species can be due to mutations in major genes. There can be due to additive effect of several genes. These genes can affect the phenotypes. We can observe qualitative changes when the mutations occur.

In a same gene, there will be different mutations. These mutations can confer a variable level of resistance towards fungicides. Due to the influence of these mutations, there will be multi-allelic resistance. The resistance obtained due to multiple alleles can be investigated with modern sequencing tools. These multiple alleles can be partially recessive and partially dominant. When the alleles which are mutant and wild are present in the fungal cells, we can say that the fungal members are resistant towards fungicides. The polygenic resistance can occur due to presence of mutations in the genes. We cannot detect the polygenic resistance in the field. In case of B.graminis, there are some reports of occurrence of polygenic inheritance. Nectriahaematococca has developed resistance towards dodine which is due to the onset of polygenic inheritance. The resistance towards fenarimol was developed in N.haematococca. This species also developed resistance to terbinafine and fenpropimorph. The cytoplasmic genes are located in the plasmids, mitochondria and some viruses.

The genes which are needed for fungi to generate resistance come from the viruses, plasmids and episomes. Variation in the level of ploidy can influence the alleles which are located on the locus. These genes can participate in producing resistance. The pathogenic fungi can be in the haploid stages most of the time. In case of fungi which are in haploid condition, the mutations can occur immediately. It can be determined that the resistance generating mutations can exist in haploids. There will be many nuclei in the dikaryons and heterokaryons. Due to the presence of many nuclei there will be genetic complementation. The heterokaryons can have nuclei which carry both sensitive and resistant genes. In this case the fungal members can grow irrespective of the fungicide application. The dominance and resistance can be investigated with the help of hyphal anastomosis.

The resistance level can be calculated with the aid of dose-response curves. The small variations can decide if resistance occurs due to polygenes or a single gene. The mutant genes which can result in the fungal resistance can also provide resistance to other fungicides. The mutations which can confer resistance can influence many characters such as successful reproduction, survival, virulence and competitiveness. The mechanism which can initiate resistance in the fungal members can be linked with the host species also. The mutations can be produced by treating hyphae and conidia with chemicals and physical mutagens. The physical mutagens can bring about lethality atleast in selective media. The genetic variation can be studied with the help of restriction fragment length polymorphisms or RFLPs. Some of the other techniques which can be useful are single nucleotide polymorphisms or SNPs. Allele -specific PCR can use for that matter.

The ability of a fungal pathogen to retain tolerance towards fungicides can be defined as fitness of that fungal species. We can define the reproductive success and survival of any allele or group or even individual as fitness. The resistance towards DMI was more distinct. The DMI fungicides can be crucial in regulating the scab on the pear and apple. Its application was reduced to two times or three times in one year. This is because the fungal species developed resistance to DMI fungicide. Succinate dehydrogenase inhibitor or SDHI fungicide was used to control fungi. The fungi developed resistance towards this fungicide as it acts on a single site. Due to problems associated with the SDHI compounds, their usage was stopped. Tradition fungicides such as chlorothalonil and albesilate were used.

The resistance management methods can include stopping of curative fungicides, fungal alternation, and usage of fungal mixtures and spray management. The resistance offered by the fungal members can be attributed to a single point mutation. The best example to explain the single point mutations are QoI resistance. Due to the occurrence of a single mutation, there will be protection from the fungicides. In some cases, there will be multiple mutations which can be stored in a gene. This will create resistance. When there is a mutation, there will be fungal resistance, but the same mutation can disturb the biochemical and physiological processes. The fitness of the fungal pathogens can be compromised due to the resistance mutations. It is important to calculate the fitness of resistant and sensitive pathogen strains. The evaluation must be done in an environment that is non-selective.

The mechanism of the fungal resistance towards fungicides must be studied so that the selection of fungicides can be decided. The microtubule filaments comprise a protein dimer called tubulin. The tubulin molecule consists of alpha and beta subunits. This is crucial in the division of the nucleus, especially in the eukaryotic cells. The fungicides used in agriculture should act on the tubulin molecules. These fungicides can be classified into three main classes. They are benzamides, N-phenylcarbamates and benzimidazoles. The pro-fungicides can be converted into carbendazim. Some of the anti-tubulin fungicides are zarilamide, ethaboxam, zoxamide, diethofencarb and benomyl. N-phenylcarbamate herbicides can act on the fungal species which are resistant to the benzimidazoles.

The succinate dehydrogenase inhibitors or SDHIs are considered an important group of fungicides. These chemicals can treat several diseases. Some of the SDHIs are sodaxane, penthiopyrad, isopyrazam and others. These compounds can target the respiratory chain. Due to its effectivity on the fungal species, the SDHIs were made complex to combat the growing demands.The fungicides which act as respiration inhibitors are considered successful in combating pathogens. The fungicides which can attack the respiration apparatus can be crucial in controlling the fungal members. The fungicides which can disrupt the respiration process can be classified into three groups based on their activity. Out of the three groups,the QoI group is important. Strobilurins belong to this group. In this group, there are 9 chemical classes that come under 20 compound classes.

Qil compounds can be divided into two important chemical groups. They are named as sulfamoyl triazoles and cyanoimidazoles. In order to regulate oomycetes, there are 16 groups that can be discussed. Some of the single- site inhibitors are carboxylic amides, quinine outside inhibitors or QoIs and phenylamides or PAs. Multisite inhibitors include phthalimides and dithiocarbamates. Some of the compounds which can be used in treating seeds are hymexazol, QoIs, Qils and Pas.

The organic and inorganic fungicides need a perfect plant health management strategy. Integrated pest management mainly involves pathogen or pest. In any plant health management system, it is important to identify plant and site on which plant grows. It is also crucial to understand the resistant varieties which are being used. The strategies which are needed to keep the plants growing in that area active can be important. Prevention can be considered as an important method in any plant health management system. Some of the organic fungicides involve bicarbonates, oils, copper and sulfur. Sulfur is considered as the oldest fungicide.

4

Insecticides

CONTENTS

4.1 INTRODUCTION

There are more than 10,000 insect species which are involved in damaging crops. Out of these, only 700 species create much damage to the crops (reference). In World War II, some of the insecticides used in the combat were DDT and other synthetic organic insecticides. Organochlorines have chlorine, hydrogen and carbon groups. They are known as chlorinated hydrocarbons or chlorinated insecticides. Diphenyl aliphatics are considered as the oldest group of the organochlorines. Some of the compounds such as methoxychlor, chlorobenzilate, ethylan, dicofol, DDD and DDT were classified as diphenyl aliphatics. DDT was discovered by Dr. Paul Muller in 1948. This insecticide was used to control malaria and several other insects. Due to its usage, several diseases such as malaria and yellow fever were controlled. DDT acts on the sodium and potassium balance in the neurons. This imbalance will be achieved by acting on the neural transmission in insects and mammals. There will be no control on the nerve impulses so the affected organisms die due to DDT jitters. DDT can work best in low temperatures and it will exhibit high rates of toxicity.

		Trade names	
Groups	Common names	Individual active ingredients	Formulated mixtures
1A	methomyl	Lannate	
1B	acephate	Acephate	
	chlorpyrifos	Lorsban Advanced, Chlorpyrifos, Govern, Hatchet, Nufos, Vulcan, Warhawk, Whirlwind, Yuma	Tundra Supreme, Cobalt, Cobalt Advanced, Stallion, Match-Up
	dimethoate	Dimethoate	
3A	alpha-cypermethrin	Fastac	
	beta-cyfluthrin	Baythroid	Leverage
	bifenthrin	Tundra, Sniper, Fanfare, Discipline, Brigade, Bifenture	Justice, Match-Up, Tundra Supreme, Brigadier, Swagger, Skyraider, Hero, Steed, Triple Crown
	cyfluthrin	Tombstone	
	deltamethrin	Delta Gold, Batallion	
	esfenvalerate	Asana XL, Adjourn	
	gamma-cyhalothrin	Declare, Proaxis,	Cobalt
	lambda-cyhalothrin	Warrior II, Grizzly Z, LambdaStar, Lambda-Cy, Lamcap, Province, Silencer VC, Taiga Z,	Besiege, Cobalt Advanced, Double Take, Endigo, Seeker
	permethrin	Arctic	
	zeta-cypermethrin	Mustang Maxx, Respect	Hero, Steed, Stallion, Triple Crown
4A	acetamiprid		Justice
	chlothianadin		Belay
	imidacloprid	Prey, Admire Pro, ADAMA Alias, Wrangler, Nuprid, Sherpa,	Leverage, Brigadier, Swagger, Skyraider, Triple Crown
	thiamethoxam		Endigo

Figure 4.1. Insecticides.

Source: https://www.researchgate.net/publication/221803317_Fungicide_impacts_on_photosynthesis_in_crop_plants/link/0fcfd50c061d089138000000/download

4.2 CLASSIFICATION

Group 1 – Acetylcholinesterase (AChE) inhibitors
Group 2 – GABA-gated chloride channel antagonists
Group 3 – Sodium channel modulators
Group 4 – Nicotinic acetylcholine receptor (nAChR) agonists
Group 5 – Nicotinic acetylcholine receptor (nAChR) allosteric activators
Group 6 – Chloride channel activators
Group 7 – Juvenile hormone mimics
Group 8*– Miscellaneous nonspecific (multisite) inhibitors
Group 9 – Modulators of chordotonal organs
Group 10 – Mite growth inhibitors
Group 11 – Microbial disruptors of insect midgut membranes
Group 12 – Inhibitors of mitochondrial ATP synthase
Group 13* – Uncouplers of oxidative phosphorylation via disruption of the proton gradient
Group 14 – acetylcholine receptor (nAChR) channel blockers
Group 15 – Inhibitors of chitin biosynthesis, type 0
Group 16 – Inhibitors of chitin biosynthesis, type 1
Group 17 – Moulting disruptor, Dipteran
Group 18 – Ecdysone receptor agonists
Group 19 – Octopamine receptor agonists
Group 20 – Mitochondrial complex III electron transport inhibitors
Group 21 – Mitochondrial complex I electron transport inhibitors
Group 22 – Voltage-dependent sodium channel blockers
Group 23 – Inhibitors of acetyl CoA carboxylase
Group 24 – Mitochondrial complex IV electron transport inhibitors
Group 25 – Mitochondrial complex II electron transport inhibitors
Group 28 – Ryanodine receptor modulators
Group UN*– Compounds of unknown or uncertain MoA2

Figure 4.2 Classification of insecticides.

Source:https://www.novasource.com/en/resources/insecticide-classification-guide

IRAC GROUP	MODE OF ACTION	CHEMICAL FAMILY (GROUP)	ACTIVE INGREDIENTS
IA	Acetylcholine esterase inhibitors	Carbamates	Aldicarb, Alanyearb, Bendiccarb, Benfuraearb, Butocatboxim, Butoxycarboxim, Carbanl, Carbofuran, Carbosulfan, Ethrokncarb, Fenobuurb, Fonnaanate, Furathiccarb, Isoprocarb, Methiocarb, Methomyl, Ski°loath, Oxamyl, Pirimicarb, Propoxur, Thiodicarb, Thiofanox, Trimethacarb, XMC, Xylylearb
IA		Triazemate	Triazanate
I B		Organophosphates	Acephate, Azamethiphos, Azinphos-ethyl, Azinphosmethyl, Cadusafos, Chlorethoxyfos, Chlorfenvinphos, Chlonnephos, Chlotpyrifos, Chlorpyrifosmethyl, Coumaphos, Cyanophos, Demcton-S-methyl, Diazinon, Dichlones· DDVP, Dicrotophos, Dimcihoate, Dimethylvinphos, Disulfoton, EPN, Ethion, F:thoprophos, Famphur, Fenamiphos, Fenitrothion, Fenthion, Fosthiazate, lieptenophos, Isoknphos, Isopropyl 0- methoxyaminothio=phosphorylt salicylatc, Isoxathion, Malathion, Mecarbam, Methamidophos, Methidathion, Mcvinphos, Monocrotophos, Nakd, ()methodic, Oxydemeton-methyl, Parathion, Parathion-methyl, Phenthoate, Phonic, Phosalone, Phosmet, Phosphamidon, Phoxim, Pirimiphos-, ethyl, Profatofos, Propetamphos, Prothiolbs, Pyraclofos, Pyridaphenthion, Quin:110os. Sullotep. Tebupirimfos, Temephos, Terbufos, Tetrachloninphos, Thiometon, Triazophos, Triehlorfon, Vamidothion

Figure 4.3. Modes of action (Group 1A to 1B).

Source: http://pested.okstate.edu/pdf/insecticide%20moa.pdf

IRAC GROUP	MODE OF ACTION	CHEMICAL FAMILY (GROUP)	ACTIVE INGREDIENTS
2A	GABA-gated chloride channel antagonists	Cyclodiene organochlorines	Chlordane, Endosulfan, gamma-HCH (Lindane)
2B		Phenylpyrazoles (Fiproles)	Ethiprole, Fipronil
3	Sodium channel modulators	DDT	DDT
3		Methoxychlor	Methoxychlor
3		Pyrethroids	Acrinathrin, Allethrin, d-cis-trans Allethrin, d-trans Allethrin, Bifenthrin, Bioallethrin, Bioallethrin S-cylclopentenyl , Bioresmethrin, Cycloprothrin, Cyfluthrin, beta-Cyfluthrin, Cyhalothrin, lambda-Cyhalothrin, gamma-Cyhalothrin, Cypermethrin, alpha-Cypermethrin, beta-Cypermethrin, theta-cypermethrin, zeta-Cypermethrin, Cyphenothrin , (1R)-trans- isomers], Deltamethrin, Empenthrin , (EZ)- (1R)- isomers], Esfenvalerate, Etofenprox, Fenpropathrin, Fenvalerate, Flucythrinate, Flumethrin, tau-Fluvalinate, Halfenprox, Imiprothrin, Permethrin, Phenothrin [(1R)-trans-isomer], Prallethrin, Resmethrin, RU 15525, Silafluofen, Tefluthrin, Tetramethrin, Tetramethrin [(1R)-isomers], Tralomethrin, Transfluthrin, ZXI 8901
3		Pyrethrins	Pyrethrins (pyrethrum)
4A	Nicotinic Acetylcholine receptor agonists / antagonists	Neonicotinoids	Acetamiprid, Clothianidin, Dinotefuran, Imidacloprid, Nitenpyram, Thiacloprid, Thiamethoxam
4B		Nicotine	Nicotine
4C		Bensultap	Bensultap
4C		Cartap hydrochloride	Cartap hydrochloride
4C		Nereistoxin analogues	Thiocyclam, Thiosultap-sodium
5	Nicotinic Acetylcholine receptor agonists (allosteric) (not group 4)	Spinosyns	Spinosad
6	Chloride channel activators	Avermectins, Milbemycins	Abamectin, Emamectin benzoate, Milbemectin
7A	Juvenile hormone mimics	Juvenile hormone analogues	Hydroprene, Kinoprene, Methoprene
7B		Fenoxycarb	Fenoxycarb
7C		Pyriproxyfen	Pyriproxyfen

Figure 4.4. Modes of action (Group 2A to 7C).

Source: http://pested.okstate.edu/pdf/insecticide%20moa.pdf

IRAC GROUP	MODE OF ACTION	CHEMICAL FAMILY (GROUP)	ACTIVE INGREDIENTS
8A	Compounds of unknown or non-specific mode of action (fumigants)	Alkyl halides	Methyl bromide and other alkyl halides
8B		Chloropicrin	Chloropicrin
8C		Sulfuryl fluoride	Sulfuryl fluoride
9A	Compounds of unknown or non-specific mode of action (selective feeding blockers)	Cryolite	Cryolite
9B		Pymetrozine	Pymetrozine
9C		Flonicamid	Flonicamid
10A	Compounds of unknown or non-specific mode of action (mite growth inhibitors)	Clofentezine	Clofentezine
10A		Hexythiazox	Hexythiazox
10B		Etoxazole	Etoxazole
11A1	Microbial disruptors of insect midgut membranes (includes transgenic crops expressing Bacillus thuringiensis toxins)	B.t. subsp. israelensis	Bacillus thuringiensis subsp. Israelensis
11A2		B. sphaericus	Bacillus sphaericus
11B1		B.t. subsp. Aizawai	Bacillus thuringiensis subsp. Aizawai
11B2		B.t. subsp. kurstaki	Bacillus thuringiensis subsp. Kurstaki
11C		B.t. subsp. tenebrionis	Bacillus thuringiensis subsp. Tenebrionis
12A	Inhibitors of oxidative phosphorylation, disruptors of ATP formation (inhibitors of ATP synthase)	Diafenthiuron	Diafenthiuron
12B		Organotin miticides	Azocyclotin, Cyhexatin, Fenbutatin oxide
12C		Propargite	Propargite
12C		Tetradifon	Tetradifon
13	Uncouplers of oxidative phosphorylation via disruption of proton gradient	Chlorfenapyr	Chlorfenapyr
13		DNOC	DNOC
15	Inhibitors of chitin biosynthesis, type 0, Lepidopteran	Benzoylureas	Bistrifluron, Chlofluazuron, Diflubenzuron, Flucycloxuron, Flufenoxuron, Hexaflumuron, Lufenuron, Novaluron, Noviflumuron, Teflubenzuron, Triflumuron
16	Inhibitors of chitin biosynthesis, type 1, Homopteran	Buprofezin	Buprofezin
17	Moulting disruptor, Dipteran	Cyromazine	Cyromazine

Figure 4.5. Modes of action (Group 8A to 17).

Source: http://pested.okstate.edu/pdf/insecticide%20moa.pdf

IRAC GROUP	MODE OF ACTION	CHEMICAL FAMILY (GROUP)	ACTIVE LNGREDIENTS
18A	Ecdysone agonists / moulting disnqxors	DiacyThydrazines	Chrounfenozide. Halofenozide. Nlethoxyfenozide. Tebufenozide
18B		Azadirachtin	Azadirachtin
19	°**dopamine*** agooists	minx	Am itraz
20A	Mitochondria' complex III election transpon inhibitors (Coupling site II)	Hydramethylnon	Hydramelninon
2013		Acequinacyl	Adquinocyl
20C		Fluacryprim	Fluacryprim
21	Mizochondrial compkx I electron transport inhibitors	METI acaricides	Fenanquin. Fenpyroximate. Primidifen Pridaben Tebufenpyrad. Tolftanpyrad
21		Rotenone	Rotenone
22	Voltage-dependent sodium channel block's	Indoxacarb	Indoxacalb
23	Inhibitors of lipid synthesis	Tetronic acid derivatives	Spirodiciofen. Spiromesifen
24A	Mitochondrial complex IV electron transport inhibitors	Aluminium phosphide	Aluminium phosphide
24B		Cyanide	Cyanide
24C		Phosphine	Phosphine
25	Neuronal inhibitors (unknoun mode of action)	Bifenazate	Bifenazate
26	Aconitase inhibitors	Fluoroacetate	Fluoroacetate
27A	Synergists	P450-dependent monooxygenase inhibitors	Piperonyl butoxide
27B		Esterase inhibitors	Tribufos (DEF)
28	Ryanodine receptor modulators	Flubendiamide	Flubendiamide

UNA	Compounds with unknown mode of anion˙	Benzoximate	Benzoximate
UN13		Chinomethionat	Chinomethionat
UNC		Dicofol	Dicofol
UND		Pridalyl	Pridalyl
NSA	Miscellaneous non-specific (multi-site) inhibitors	Borax	Borax
NSB		Tartar emetic	Tartar emetic

Figure 4.6 Modes of action (Group 18A to NSB).

Source: http://pested.okstate.edu/pdf/insecticide%20moa.pdf

Hexachlorocyclohexane is called as HCH. It is also known as benzenehexachloride or BHC. This chemical can exist in five isomers. The gamma isomer of this chemical was produced on an industrial scale in the name of lindane. It is flavorless and odorless. The cost of HCH is low. This chemical can enhance the rate of respiration in insects. It can work as a neurotoxicant. Post World War II, cyclodienes were produced. The important cyclodienes were dieldrin, aldrin and chlordane. Others which came later are heptachlor, endrin, mirex and chlordecone. These chemicals can have stability in the soils and they are less reactive to the UV light. They can be stable in the soils and so they are being used as soil insecticides all over the world. Some of the cyclodienes which are used even today are dieldrin, aldrin, hepatochlor and chlordane. They are used to control several insects and termites. They mainly act on the larval stages. The activity of the cyclodiene will get increased with temperature and so we can describe them as exhibiting positive temperature correlation. These insecticides can act on gama-aminobutyric acid receptor or GABA receptor. There will be an inhibitory mechanism. Due to these chemicals, the ion permeability of the chloride ion will get increased. This occurs in the neurons and so the chloride ions may not reach the neurons and hence the calming effects offered by GABA will be halted. In 1947, polychloroterpenes were discovered.Another chemical which acts as an insecticide is strobane.Toxaphene was first used to control insect pests on cotton crops. DDT resistance in the insects was observed in 1965 and since then, toxaphene was used as organophosphate insecticide. Toxaphene was prepared by mixing 10C polychlorinated compounds. This chemical can survive in the soils and offers protection for longer periods. It has higher metabolic rates in the birds and mammals and hence less toxic to them. It can kill fishes and so was banned in 1983.

All the insecticides which contain phosphorous can be called as organophosphates or OPs. They are toxic to vertebrates. In World War II, nerve gases named tabun, soman and sarin were tested. Nicotine was used as insecticide then, but due to its unavailability, there was a search towards other chemical substitutes. The OPs exhibit a high rate of toxicity to the vertebrates than other insecticides. They are unstable and may not survive for longer periods. They inhibit cholinesterase or ChE which is an important enzyme in the nervous system. There will be phosphorylation of the enzymes. Due to this, the acetylcholine or Ach will get accumulated and so there will be paralysis in the insects. The OPs can be classified into 6 sub-classes. They are phosphoramidates, phosphorothiolates, phosphorodithioates, phosphorothioates, phosphonates and phosphates. The OPs are classified into aliphatic, phenyl and heterocyclic derivatives. The OPs which are aliphatic are chain- like compounds. Some of the other prominent examples are malathion and trichlorfon. The OPs which are having phenyl rings can be considered stable. Some of the examples of these compounds are parathion and sulprofos. Diazinon is a heterocyclic compound which has insecticidal activity.

The organosulfurs can exhibit lower toxicity on the insects. They can work as acaricides or miticides. In these compounds, we can observe 2 phenyl rings. Almost all the chemical structure is the same as DDT except that sulfur replaces the central atom. Some of the examples of the organosulfurs are ovex, comite and tetradifon. The carbamate compounds or their derivatives can work as insecticides. They act on the cholinesterase. Carbaryl compound named Sevin was first produced in 1956. Some of the other carbamates that can work as insecticides are aldoxycarb, fenoxycarb and carbofuran. The carbamates can interrupt cholinesterase. Some of the compounds can inhibit aliesterase.

Formamidines are insecticides which constitute a small group. Ovasyn, amitraz and fundal are some of the examples of formamidines. They exhibit clear symptoms than other insecticides. They can inhibit monoamine oxidase. Serotonin and norepinephrine are considered as neurotransmitters. Because of the disruption of these systems, there will be deposition of these compounds at alarming levels. These compounds can be called as biogenic amines. These compounds can damage insects. DInitrophenols can exhibit a high rate of toxicity than fungicides and herbicides. Dinocarp can be considered as an efficient miticide. They mainly inhibit oxidative phosphorylation. Due to this inhibition, ATP production will be hindered. Organotonins canbe classified as acaricides. They are more toxic than fungicides. Cyhexatin is an important acaricide. Vendex is effective on ornamentals, greenhouse crops, citrus fruits and other plants. The compounds made up of tin can inactive the process of oxidative phosphorylation. Trialkyl tins can impair the photophosphorylation process in the chloroplasts. They can work as algicides.

Pyrethroids are powerful insecticides. Pyrethrum is a naturally occurring compound that is weak and maybe inactivated due to sunlight. Several analogs of these compounds were synthesized. They are called synthetic pyrethroids. First- generation pyrethroids include allethrin. The second- generation pyrethroids comprise of tetramethrin. The third generation comprises of fenvalerate. These compounds are stable and exhibit photostability. The nicotinoids can be classified as insecticides. They are called neonicotinoids and chloronicotinyls. Imidacloprid can be classified as a systemic insecticide. We can use this insecticide on the leaves, seeds and soils. This chemical can control potato beetles. Some of the other nicotinoids are nitenpyram and platinum. These insecticides can influence the central nervous systems. They alter the postsynaptic receptors. These receptors are called as nicotinic aceylcholine receptors.

Spinosyns can be considered as the latest class of insecticides. The market name for this chemical is spinosad. It is a metabolite derived from Saccharopolyspora spinosa which is classified as actinomycetes. This insecticide acts on the nicotinic acetylcholine receptors. Fiprolesiscalled as phenylpyrazoles which can act as powerful insecticides. This insecticide has both stomach and contact activity. We can control both foliar and soil insects such as rice water weevil, potato beetle and corn rootworm. These insecticides can control termites, ants and cockroaches. The insects which are resistant to carbamate, organophosphate and pyrethroid can be treated with fipronil. This insecticide can alter the GABA mediated channels in the neurons. The calming effects produced by the GABA molecules will be halted. Pyrroles are important insecticides which are to be considered while studying crop sciences. Chlorfenapyr can be classified under pyrroles. They work as stomach insecticide and contact insecticide. This chemical can act as a miticide. We can use this chemical on cotton crops and soybean. Several ornamental plants can be protected with this insecticide. Some of the insects which can be controlled with this insecticide are potato beetle, aphids, leafminers, mites and caterpillars. Chlorfenapyr can inhibit oxidative phosphorylation and ATP production will be ceased. This will kill insects.

Pyrazoles constitute fenpyroximate and tebufenpyrad. There are classified as stomach miticides and non- systemic insecticides. They are active onthrips, whitefly, aphids and psylla. This chemical can be used on vegetables and soybeans. These chemicals can control all the life stages of the insecticides. It inhibits the production of ATP. Pyridazinones are crucial in controlling insects. Pyridaben can be classified as pyridazinone. This chemical can disrupt the electron transport site of the insects. The quinazolines can be used in the form of fenazaquin. It can be classified as a stomach miticide and contact insecticide. It controls all the life stages of the insects. It can inhibit the electron transport process. Benzoylureas can be classified as insecticides that can act on the growth regulators or IGRs. They can interrupt the nervous system in insects. With this insecticide, we can control beetle larvae and caterpillars. Triazine can be considered as an inhibitor of chitin synthesis. It can control dipterous species. Most of the leafminers can be regulated with these insecticides. These insecticides can be used in the mushroom houses. They act on all the larval stages of the insects. These chemicals can inhibit the chitin synthesis and hence there will be imperfect growth stages which can lead to the death of the insects in their larval stages. The cuticle will be irregular and hence the growth patterns will be altered. The botanical insecticides can be considered as natural insects. They are practically derived from plants.

Turpentine, camphor and hellebore can be considered as important plant products that can be treated as botanicals. Pyrethrum can be isolated from the flowers. We can extract these compounds from the chrysanthemum flowers. We can use it safely and it has no other problems. Pyrethrum can be considered as a mix of 4 compounds. They are cinerins I and II, pyrethrin I and II. This chemical can be considered as axonic poison. They can be called as synthetic pyrethroids and DDT. Pyrethroids and pyrethrum can be considered as having insecticidal effects. They can influence the nervous system of insects. The ganglia and the synapses will be damaged.

Nicotine can be isolated from the tobacco plant through several methods. These chemicals can kill caterpillars and aphids. It is an alkaloid that can be useful in disrupting the physiological properties of the insects. Some of the alkaloids are quinine and caffeine. Nicotine can inflict the same response as acetylcholine. Due to the activity of these compounds, there will be death to the insects. Rotenoids or rotenones can be isolated from the roots. We have two families that can help in the production of these insecticides. The families are lonchocarpus and derris. It can kill fishes. Rotenone can act as an inhibitor of enzymes involved in respiration. Limonene is considered as botanicals. They can be isolated from the peels of the citrus fruits. These chemicals can act on the ticks, lice and pets. Limonene can comprise 98% of the peel of the oranges. The action is similar to the pyrethrum. It affects the peripheral nervous system. It cannot inhibit cholinesterase inhibitor. Neem oil can be obtained by squeezing the seeds of the neem tree. Azadirachtin is an important ingredient of the neem oil. Azatin can be used as insect growth regulators. Azadirachtin can inhibit the production of ecdysone. This hormone helps in the molting process.

The activators or synergists can be toxic in nature and they can work as insecticidal compounds. They can be used to enhance the activity of the insecticides. The synergists can lower the activity of cytochrome p 450 PSMOs or polysubstrate monooxygenases. Microsomes can be interrupted due to these insecticides. These enzymes can degrade foreign materials. The synergists can associate with the oxidation enzymes. They can help the toxicant from being degraded. The antibiotics include avermectins. This type of chemicals can act as acaricidal, insecticidal and antihelminthic chemicals. They are produced from Streptomyces avermitilisbacteria.

They can kill the parasites and several caterpillars. Larvae will get paralyzed due to the application of these chemicals. These chemicals can damage the GABA molecules which are considered as neurotransmitters. The nervous system in the insects and mites can be damaged due to altered neuromuscular junction. The egg -laying capacity and the feeding systems of the larvae will be disturbed. Death is imminent on exposure to these chemicals.

Fumigants are considered crucial in crop production as their activity is high compared to other modes of applications. The fumigants can exist as gases at normal temperatures. They are organic molecules that are small and volatile. The chemical structure of these compounds is made up of F, Br or Cl. They are capable of penetratinga large number of weeds. They can control nematodes; insects and can alter the egg -laying process by involving in the chitin production process. They can kill the microbes in the greenhouse, soils and warehouses. Most commonly used fumigants are made up of methyl bromide. This chemical can be used on a wide variety of crops. Some of the chemicals which can be used as fumigants are hydrogen cyanide, ethylene chloride, ethylene oxide and others. We need to use the magnesium phosphide pellets which have a tendency to react with the air and there will be the production of gas which is of higher density. The fumigants can work as narcotics. They can be soluble in fat and so we call them as liposoluble. They produce sleep and narcosis in the insects.

The insect repellants can be used to control insects on a large scale. We can use smoke as an insect repellant. Some of the repellants which can be used are indalone and dimethyl phthalate. The inorganic insecticides don't contain carbon. They resemble salts. An important property of these repellants is that they cannot be vaporized easily. Sulfur can be considered as an effective insect repellant. The sprays and powders produced from the sulfur molecules can be used to control insects. Some of the inorganic compounds which can be used as inorganic insecticides are made up of fluoride, selenium, antimony, arsenic and boron. The arsenic molecules can be used in the form of calcium arsenate and lead arsenate. They can disrupt the oxidative phosphorylation process. Some of the other compounds which can be used as insecticides are cryolite, sodium silicofluoride, barium fluosilicate and sodium fluoride. The fluoride ions which are present in these insecticides can damage several enzymes which comprise magnesium, calcium and iron. Some of these enzymes which can be damaged are phosphorylases and phosphatases. We can use boric acid to control insecticides. It can work on the stomach and cuticle of the insects. There will be rapid dehydration due to the absorption of the fumigants.

Some of the miscellaneous compounds can act as insecticides. The compounds such as buprofezin and pyridine can serve as efficient insecticides. They can control the growth of the whiteflies. These whiteflies have created a lot of problems in the cotton- producing areas. Tetrazines can be used in regulating the insects which can feed on the ornamentals, cotton and fruits. Enzone can be used on the citrus and grapes and control the insects which feed on them. These compounds can be effective on the soil insects and nematodes. Clandosan is the chemical derived from the shrimp shells and crab. This compound can be effectively used as a nematicide. It can be used to control nematodes and other microbes.

4.3 TOXICOLOGY

The organochlorine insecticides can be used in various malaria control programs and agricultural practices. The insecticides which are made up of chlorinated hydrocarbons canbe classified into 5 groups. They are named as DDT,toxaphene, cyclodienes and lindanes. Some of the structures which resemble cage are chlordecone and mirex. The insecticides can vary in their toxicity. Endrin can be considered toxic than the derivatives of the dieldrin or dieldrin itself.

Rising Toxicity of pesticides 1945-2003

Pesticide	Brand Name	Use	LD50 (ng/bee)	Toxicity DDT = 1
DDT	Dinocide	Insecticide	27,000.0	1
Amitraz	Apivar	Acaricide	12,000	2
Coumafos	Perizin	Acaricide	3,000	9
Taufluvalinate	Apistan	Acaricide	2,000	14
Metiocarb	Mesurol	Insecticide	230	117
Carbofuran	Curater	Insecticide	160	169
Lambda-cyhalothrin	Karate	Insecticide	38	711
Deltamethrin	Decis	Insecticide	10	2,700
Thiametoxam	Cruiser	Insecticide	5	5,400
Fipronil	Regent	Insecticide	4.2	6,429
Clothianidin	Poncho	Insecticide	4.0	6,750
Imidacloprid	Gaucho	Insecticide	3.7	7,297

Source: Dr. J.M. Bonmatin, CNRS (France)

Figure 4.7. Increasing toxicity of pesticides.

Source: Dr. J.M. Bonmatin, CNRS(France)

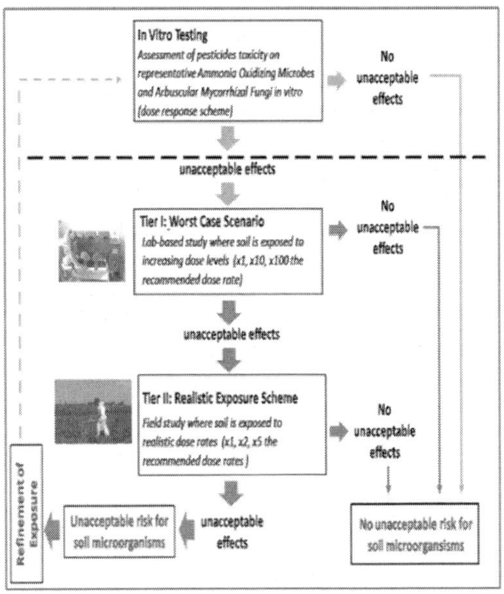

Figure 4.8 Effects of toxicity

Source:http://plantenvlab.bio.uth.gr/wp-content/uploads/2016/12/Activity-4-Toxicity-Pesticides-1024x761.jpg

The organochlorine based insecticides can produce neuronal hyperactivity. The chemicals such as toxaphene, endrin, dieldrin and aldrin can be toxic to the insects. The first symptom which can be observed in the insects after spraying these insecticides is injury. The concentration of these insecticides can be based on the concentration of the insecticides at the site of action. These chemicals can be absorbed via the skin and other organ systems such as oral and respiratory routes. Methoxychlor can exhibit minimum toxicity. DDT can exhibit minimal absorption than other chemicals. The absorption of these insects is based on the fat and fiber content of the diet and overall food intake of the insects which is also crucial to determine the insecticidal use.

These insecticides can also harm birds, animals and humans. These tissues get stored in the fat tissues. These insecticides can influence the cell membranes of the nervous system. There will be a change in the dynamics of the sodium and potassium ions. The flow of these ions will be altered and so the membranes will be damaged. Because of the interruption of the calcium ions, the ATPase activity will be influenced. The biogenic amines can produce toxicity in the chlorinated hydrocarbons. The organochlorine based insecticides and DDT can cause liver damage in rodents and human organs. These chemicals can produce tumors and so we can say that these chemicals can act as carcinogens. DDT can have the same activity as compounds such as toxaphene, dieldrin and chlordane. These compounds can also disrupt the endocrine activities of the animal systems. Due to the usage of the organochlorine pesticides, there will be chlorine oxidative stress.

There will be an onset of chronic and acute neurotoxicity in humans post -exposure to these chlorinated insecticides. These insecticides can make their way in the mother's milk and hence damage the tissue of the baby vulnerable to toxic effects. HCH or hexachlorocyclohexane can exist as 8 steric isomers. The compound lindane can be available in several formulations such as wood preservatives, wettable powders, sprays, aerosols and other concentrates. HCH molecules can control or kill the soil pests, wireworms, insects, rice insects, cotton insects and grasshoppers. We can use this chemical to protect seeds from insects. In humans, lindane is being used as pediculocide and scabicide. We will use this chemical in the form of shampoos, creams and lotions. We can observe acute poisoning as lindane can impair the rates of respiration.

5

Nematicides

CONTENTS

5.1 INTRODUCTION

The nematodes are invertebrates which are mostly non-segmented. Their body exhibits bilateral symmetryingThey hveabody cavity and a well developed digestive system. They don't have circulatory and respiratory systems. The body wall is made up of musculature, cords which are longitudinal and multiple cuticles. Close to the pharynx, ill beisthe presence of the nerve ring. The excretory system will eliminate all the w. Well The well defined osmoregulatory system is seen in these nematodes. These systems are based on the species and the developmental stages they are in. The whole body of the nematodes is based on reproductive and digestive systems. Most of the species which are classified as nematodes can be classified as free living. Methylene bromide is considered as fumigant which can be toxic to several nematodes. Chloropicrin is considered as one of the oldest fumigants. It can also work as a fungicide. We can observe both nematicidal and herbicidal activity. They can be toxic when used in the form of liquid formulations. Meta sodium can be considera fumigant which that can be used to regulate the nematodes. They can control all the weeds, insects and fungi. The compounds which are similar to the meta sodium derivatives can also act on the nematodes. There are some oth compounds calledas isothiocyanates which are considered toxic to the nematodes. Sodium tetrathiocarbonate can be used as soil fumigants and it is classified as a preplant soil fumigant. It can be used as a liquid formulation. We can use these chemicals in the form of drip and surface irrigation.

5.2 CLASSIFICATION

Aldicarb can be oxidised oxidized in the soils and there will be the formation of aldicarb sulfone. It can work as aoxycarb. We can use thisThis chemical use in the form of the soluble formulation. Carbofuran is another important chemical which can be active against the nematodes. It is classified as a nematicidal carbamate. Oxamyl can be classified as carbamate which can be produced in the form of granular and liquid form. The birds will get poisoned due to this chemical and die. Ethoprop is considered as nematicide and insecticide. It can be mobile in the soil. The half-life of this chemical can be based on the organic matter present in the soils. Fenamiphos is classifie systemic insecticide which that can kill nematodes. Cadusafos is classified as organophosphate which is non- systemic in nature. It can control both citrus nematodes and burrowing nematode. Radopholussimilis s calledasa burrowing nematodes and Tylenchulussemipenetrans is a citrus nematode. We can get this chemical as microencapsulated and granular formulations. Fosthiazate is classified as organophosphate nematicides which can be systemic in nature. We can use this chemical in the form of microglations. It controls meloidogyne Meloidogynespp. and Pratylenchus species. Resistance in the nematodes is common. The fumigation of the soils needs to be done before preparation. We need to till the soil so that the level of porosity will be increased. Due to the increase in the porosity, there will be the decomposition of the plant roots which exists as residual parts. The emulsifiable aniquid formulation can be used inthe drip irrigation systems. The easiest method of spraying the nematicides is the granuf the nematicides used inagricultural crops are carbamates, organophosphates and fumigants. The moent and efficacy of the nematicides is are considered important among the researchers working on the crop sciences. The distribution of the nematicides is crucial to understand the toxicity of the nematicides. It is important to establish ample contact time of the nematicides so that the actual toxicity is observed. The nematodes can exist as parasitic organisms in the plants. They are minute and aquatic in nature. They appear as roundworms. Water is essential for the movement of the nematodes to reach their target organisms. If water is present around the plants, the movement of the nematodes can be easier. Water can help the nematodes to maintain the water content in their bodies to an extent that they can survive for longer periods.

1. **Avermectin nematicides**
   abamectin
2. **Botanical nematicides**
   carvacrol
3. **Carbamate nematicides**
   benomyl
   carbofuran
   carbosulfan
   cloethocarb
   a. **oxime carbamate nematicides**
      alanycarb
      aldicarb
      aldoxycarb
      oxamyl
      tirpate

4. **Fumigant nematicides**
   carbon disulfide
   cyanogen
   1,2-dichloropropane
   1,3-dichloropropene
   dimethyl disulfide
   methyl bromide
   methyl iodide
   sodium tetrathiocarbonate
5. **Organophosphorus nematicides**
   a. **organophosphate nematicides**
      diamidafos
      fenamiphos
      fosthietan
      phosphamidon
   b. **organothiophosphate nematicides**
      cadusafos
      chlorpyrifos
      dichlofenthion
      dimethoate
      ethoprophos
      fensulfothion
      fosthiazate
      heterophos
      isamidofos
      isazofos
      phorate
      phosphocarb
      terbufos
      thionazin
      triazophos
   c. **phosphonothioate nematicides**
      imicyafos
      mecarphon

6.	**Unclassified nematicides**
	acetoprole
	benclothiaz
	chloropicrin
	cyclobutrifluram
	dazomet
	DBCP
	DCIP
	fluazaindolizine
	fluensulfone
	furfural
	metam
	methyl isothiocyanate
	tioxazafen
	xylenols

Figure 5.1. Nematicides- Classification.

Source:http://www.alanwood.net/pesticides/class_nematicides.html

The body of the nematode is elastic in nature so that if its body gets inflated, they may not die. In case water quantity is not sufficient, the body of the nematode gets dried and the body wall of the nematodes may get damaged and hence leaving the nematodes dehydrated. When water scarcity is higher, the nematodes will go for a suspended state. Once the water content is increased, they will come to their original state. Most of these nematodes can eat and fruits can be pronethe nematode infections. There will be a failure in the crops if the crops were grown on the sites where the nematode population is higher. Most of the damage inflictin on the crops is due to the root damage. Once the root gets damaged, the entire plant will die within no time. We can observe the infections caused due to the nematodes by analyzing some symptoms which are characteristic to the plants which are infected. Some of the symptoms are chlorotic spots, stunted growth and small when compared with healthier plants. The damage in the root system can be higher if the plants are grown in the sandy soils and coarse- textured soils. These soils may not hold a high rate of water.

These chemicals mainly kill nematodes and alongside several other pests which can cause damage to the crops. The nematicides can be classified based on the movement of the chemicals. The fumigants include metam sodium survive on the foliar tissues and roots of the plants. Nematodes can be the determinin fars inthe agricultural production. They can create more damage. The harvest may be affected in terms of quality and quantity. The vegetables, dimethyl dibromide and others.

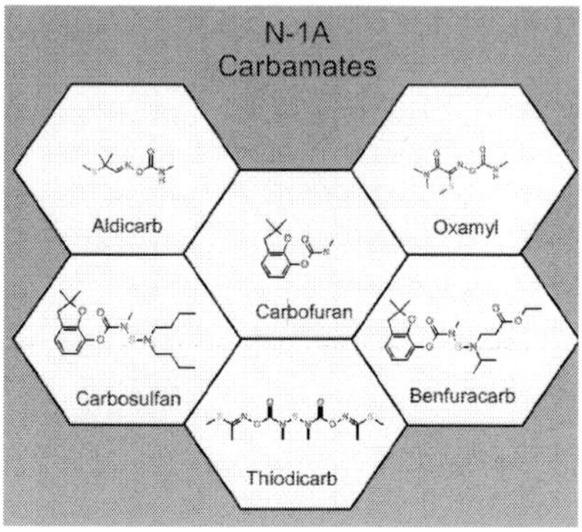

Figure 5.2. Acetylchloinesterase inhibitors (Group N-1A).

Source:https://www.irac-online.org/documents/nematicides-poster/?ext=pdf

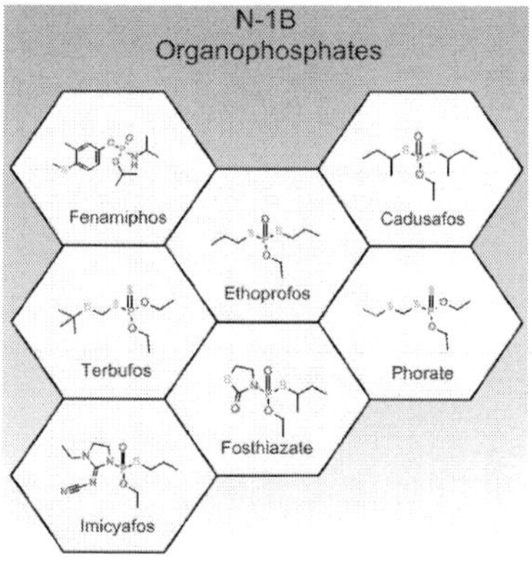

Figure 5.3. Acetylchloinesterase inhibitors (Group N-1B).

Source:https://www.irac-online.org/documents/nematicides-poster/?ext=pdf

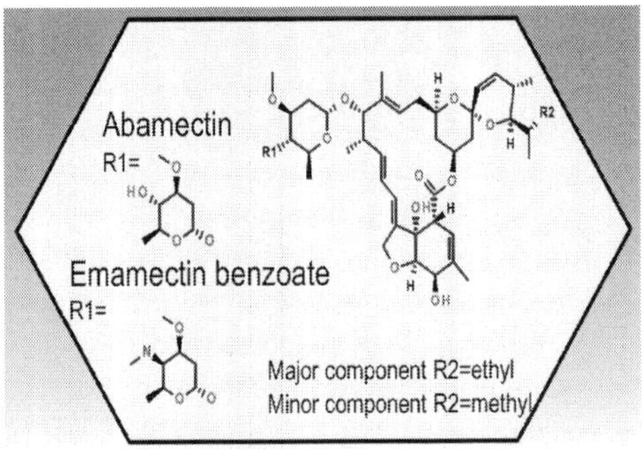

Figure 5.4. Glutamate gated chloride channel allosteric modulators (Group N-2).

Source:https://www.irac-online.org/documents/nematicides-poster/?ext=pdf

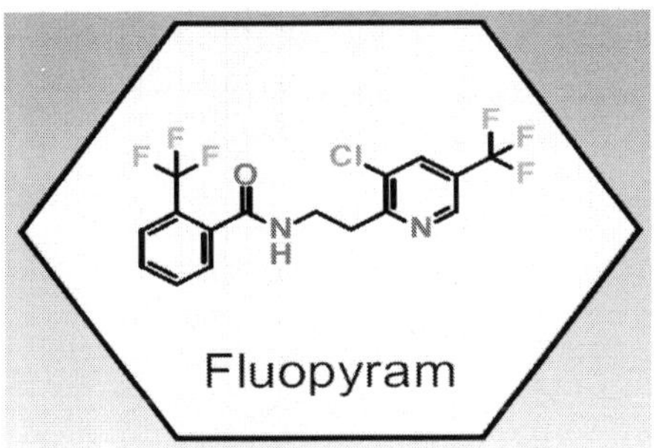

Figure 5.5. Mitochondrial complex II electron transport inhibitors (Group N-3).

Source:https://www.irac-online.org/documents/nematicides-poster/?ext=pdf

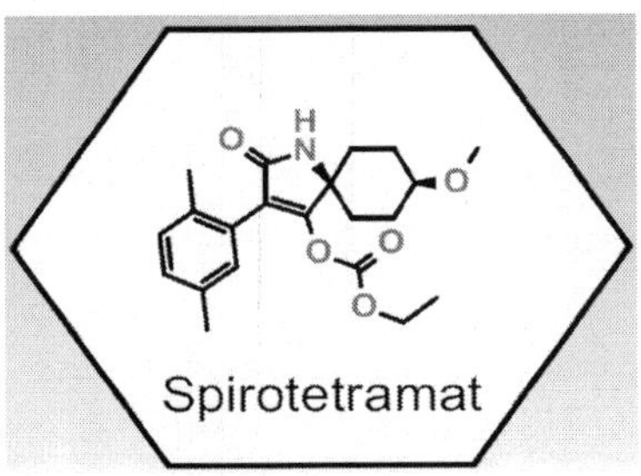

Figure 5.6. Inhibitors of acetyl CoA carboxylase (Group N-4).

Source:https://www.irac-online.org/documents/nematicides-poster/?ext=pdf

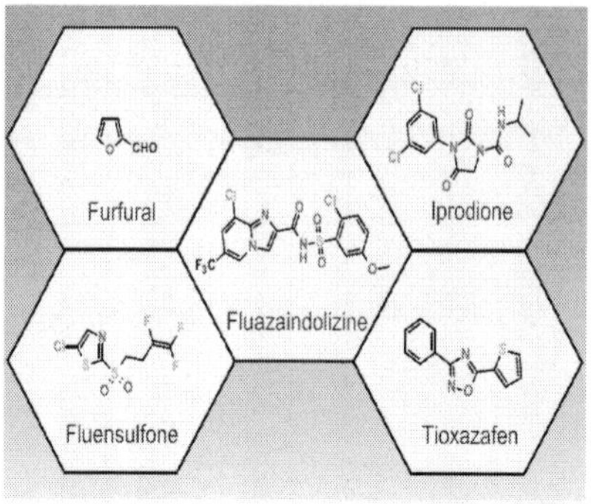

Figure 5.7. Inhibitors of acetyl CoA carboxylase (Group N-UN).

Source: https://www.irac-online.org/documents/nematicides-poster/?ext=pdf

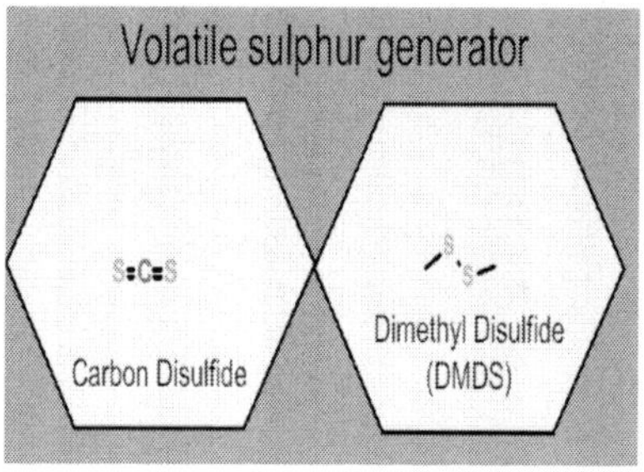

Figure 5.8. Multi-site inhibitor –volatile sulphur generator (Group N-UNX).

Source: https://www.irac-online.org/documents/nematicides-poster/?ext=pdf

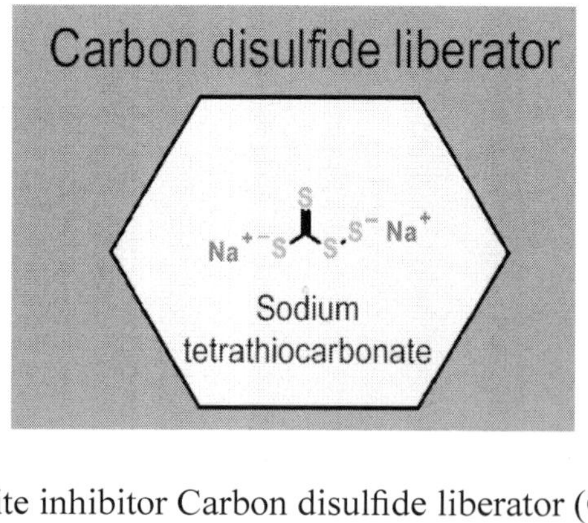

Figure 5.9. Multi-site inhibitor Carbon disulfide liberator (Group N-UNX).

Source: https://www.irac-online.org/documents/nematicides-poster/?ext=pdf

They can be vaporized and sprayed. The nematicides which are non-fumigant are Nimitz and vydate. The nematicides which are non-fumigant can be classified into systemic and contact nematicides. They can kill nematodes when comes in contact with the nematodes. They can get absorbed by the plants and then when nematodes feed on plants, they die. There will be lethal action on the nematodes which survive on the tissues. Nematicides can inflict lethal action. The fumigants can kill the nematodes instantly and there is no need of the plants to absorb the fumigants to inflict a lethal action on these species. They can affect various parts of the nematode ody an kithe tissues. TheHe vapamorthe ormetam sodium can be dissolved in water and they can get activated. The process of decomposition of these chemicals is higher in water. When the byproduct generated as a result of their activity in water reaches the body wall of the nematodes,they can affect all ts which are sensitivein nature. They can influence the respiratory systems and nervous systems.

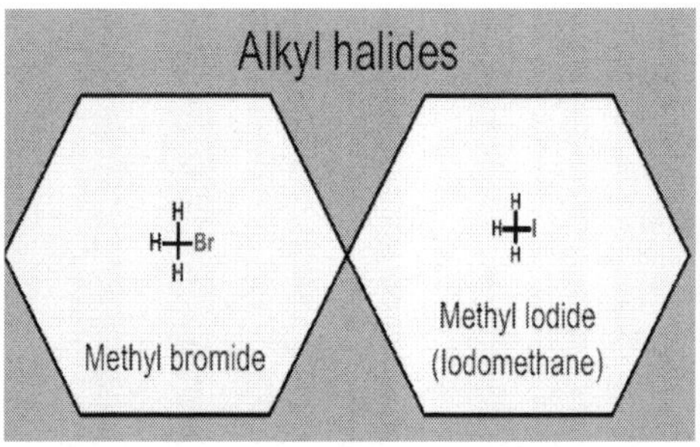

Figure 5.10. Multi-site inhibitor Alkyl halides (Group N-UNX).

Source:https://www.irac-online.org/documents/nematicides-poster/?ext=pdf

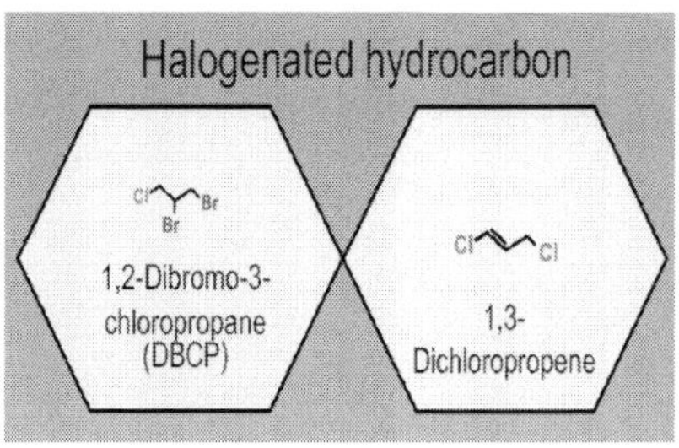

Figure 5.11. Multi-site inhibitor Halogenated hydrocarbon (Group N-UNX).

Source:https://www.irac-online.org/documents/nematicides-poster/?ext=pdf

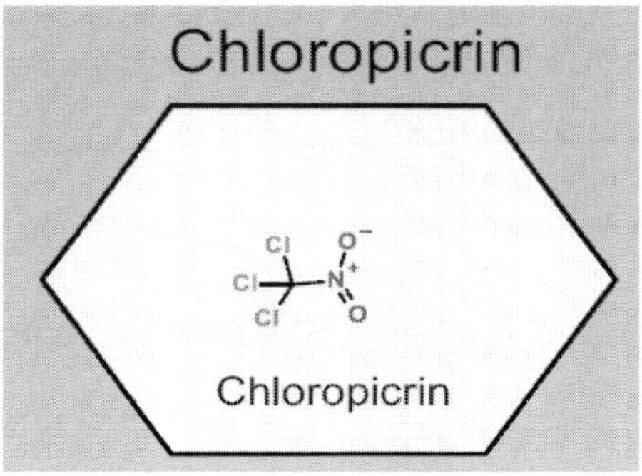

Figure 5.12. Multi-site inhibitor Chloropicrin(Group N-UNX)

Source:https://www.irac-online.org/documents/nematicides-poster/?ext=pdf

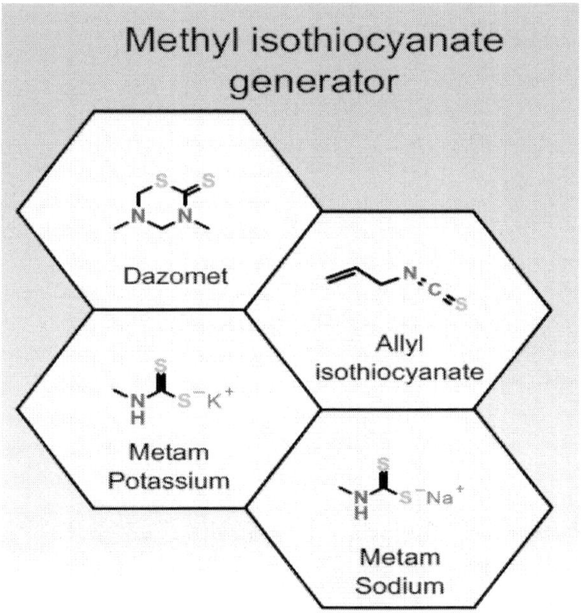

Figure 5.13. Multi-site inhibitor Methyl isothiocyanate (Group N-UNX).

Source:https://www.irac-online.org/documents/nematicides-poster/?ext=pdf

The equilibrium of the fumnts in the body of the nematodes be is reached after 4 hours or soon and it depends on the type of fumigants being used. In rare cases, the concentration of the fumigants in the nematodes can be higher than the concentration in the soils.

Some of the non- fumigants can also enter the body wall of the nematodes. Because of the carbamates and organophosphates, the nematodes die. The movement of the nematode is based on 90 neurons.

There will be neurotransmitters which include acetylcholine. Acetylcholine can be classified as an excitatory molecule.

The GABA or gamma –aminobutyric acid can be classified as an inhibitory molecule. There will be an increase in the rhythmic requency of the muscular cells due to acetylchloline.

The action of acetylcholine on the synapse junctions can be halted with the aid of the enzyme acetylcholinesterase or AChE.

Some of the inhibitors of these enzymes are organophosphate and carbamate nematicides.The plant rasite nematodes can be considered as pestich that are of major concern. They can infect the raspberry and canes.

It is impont to enhance the yield of the crops by reducingthe pests. Meloidogyne spp., can be considered as important plant pathogens whan create much damage tohe crops worldwide. This These species are calledas root-knot nematodes.

They can exist as obligate parasites. They can attack several species belonging to herbaceous, dicots and monocot plants. Itbetter practice to go for alternative methods which that can be less toxic to the environment. Salicylic acid can be used to control the defense mechanisms of the pathogens.

This chemical can also regulate insects. We can expect a high rate of reaction after 3 days. Any ingredient will be effective only if it is prepared with correct formulations. The active ingredients must reach the target site with high concentration.

The plant-parasite nematodes can be crucial in the success of any crop. Biological control of pesconsidered as critical in many crop- based industriesindustries.

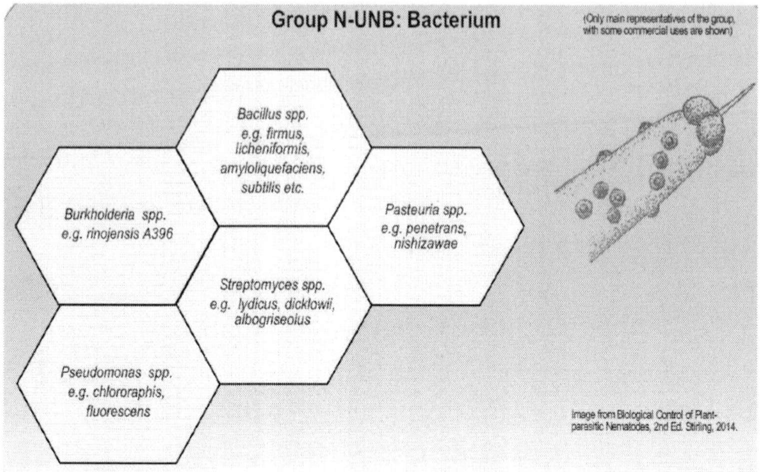

Figure 5.14. Multi-site inhibitor Bacterium (Group N-UNB).

Source:https://www.irac-online.org/documents/nematicides-poster/?ext=pdf

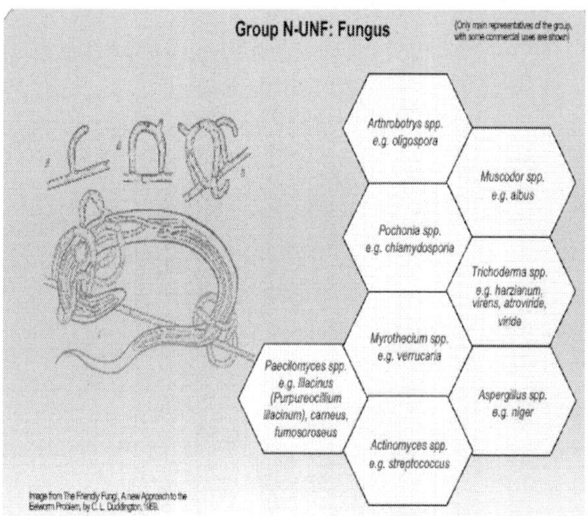

Figure 5.15. Bacterium (Group N-UNF).

Source: https://www.irac-online.org/documents/nematicides-poster/?ext=pdf

The potato cyst nematodes important plant pathogens. There two species which that attack most of the plants. TheyThose are Globoderarostochiensis and Globodera pallida. These species can lower the yields of the potato crop by 70% in some cases. Because of these pathogens, the market value of the potatoes has come down. The yield can be influenced by several factors associated with the nematode species. They are soil type, climatic conditions, potato cultivar type and sowing density. The chemical nematicides can be considered powerful than other types.

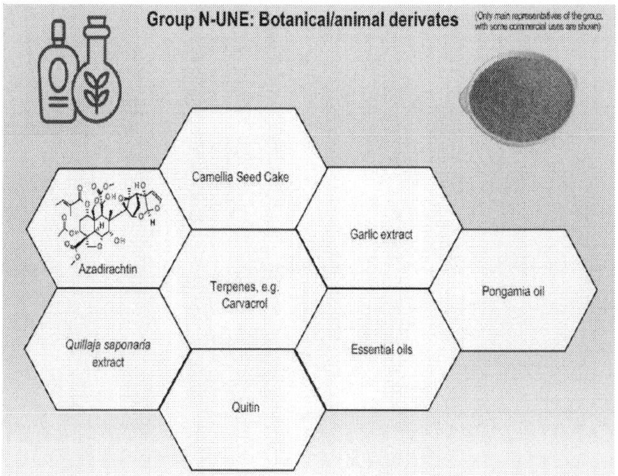

Thonematodes can be conside as destructive species which that can affect the crops at on a large scale. It is estimated that about $118 billion is being wasted early due to the attack of these phytonematodes. The phytonematodes are classified as root feeders. Most of these species complete their life cycle in the root zones and rarely can they attack the shoot zones. The root-knot nematodes are spread all over and they comprise of 150 species. Species such as Meloidogyne javanica and M.incognita are considered important. Bio-nematicides can be produced from actinomycetes, fungi, bacteria and plants. We can even use bacteria for that matter. We can use mushrooms, chitotin as bio-nematicides. Piperamides are produced by solanaceae Solanaceae members.Some of the spes such can produce these chemicals. Capsicum genus cansecrete producaicin. These chemicals can produce nematicidal properties which that can control the growth of nematodes. They can work as antifeedant, repellent and toxicant. Tagetes spp. or Marigold can produce polythienyls and polyacetylenes which can enhance the activity. Cyanogenic glycosides are produced from the euphorbiaceaeEuphorbiaceae. The leaves and roots of cassava can produce cyanogenic osides. There will be traces of lotoaustralin and linamarin in thesthese chemicals. The cells can gets damaged on exposure to these chemicals.

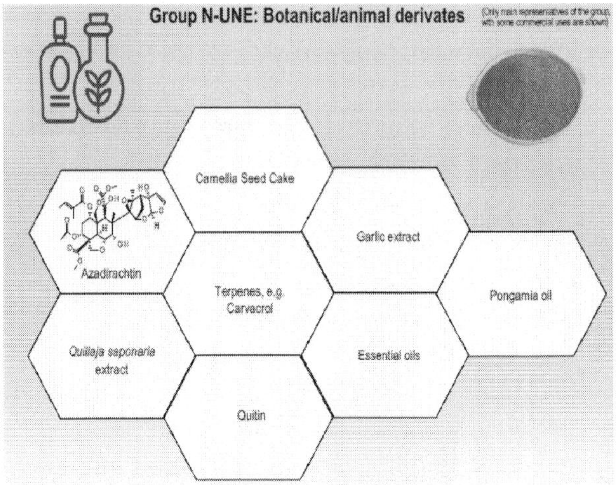

Figure 5.16. Botanical derivatives (Group N-UNE)

Source:https://www.irac-online.org/documents/nematicides-poster/?ext=pdf

The sudan grass can produce green manure and this can exhibit nematicidal properties. Thcan produce glycoside named dhurrin. This chemical can be hydrolysed hydrolyzedto produce cyanide. Terpenes can be used as nematicides. These chemicals can constitute isoprene units which are C5 compounds. These chemicals can be categorized as tetraterpenes, triterpenes, diterpenes, sesquiterpenes, monoterpens and hemi-terpenes. The hemi-terpenes are C5 compounds and tetraterpenes are C40 compounds. They can initiate paralysis in tes and other pests.

The sesquiterpenoids are produced by the malvaceae Malva. They can be nematotoxic in nature. Rishitin is produced by solanaceae Solanaceae members. This chemical can be classified as sesquiterpenoid phytoalexin. These chemicals are found in the potato tubers. These chemicals can be generated when exposed to potato rot nematodes. Odoracin is a diterpenoid which is extracted Daphne odora can produce these toxins. The cultivars which are resistantresistant cultivars and crop rotation techniques, we can control the growth of the nem. Overutilization of the nematodes has produce resistant varieties which that are hard to be controlled. It is beneficial to use the bicontrol agents to control the nematodes. All the nematode infections occurs in the rhizosphere zone. This is the area where the microbes attack. The rhizobacteria and parasitic nematodes can survive at the same niche which is called as the rhizosphere. They can exhibit symbiotic and antagonistic relationships. The rhizobacteria can control the growth of the nematodes. They can regulate twth of nematodes by releasing the nematicides. They can produce enzymes which that can degrade the cell walls of the nematodes and antibics which can check the growth of the nematodes. acteria can increasethe plant growth by producing chemicals such as ACC demainase deaminase or aminocyclopropane carboxylicid and IAA or Indole 3 acetic acid. Many volatile compounds can be produced bythe rhizobacteria. There are many enterobacter members which can grow in the root regions. They can enhance the plantgrowth and hence their yields will be enhanced proportionally. The plant growth can be enhanced by the rhizobacteria and their use as biocontrol agents is crucial. Enterobacter asburiae was isolated from the roots of the tomato plants. Foliar nematodes can be considered crucial in understanding the nematode menace. The most popular species which is considered beneficial in this case is the Aphelenchoidesfragariae. They can attack the ornamental plants which are spread over 47 plant families and 25 species. We can recognize this infection by observing the lesions generated on the leaves. Earlier, the chemical- based nematicides were used to halt the growth of the nematodes. Some of the chemicals which are used to control the nematodes are parathion, oxamyl and methyl bromide. These chemicals impose a severe threat to the environment and so their usage was stopped.

1,3-dichloropropene was successfully tested as a nticide as it won't contaminate the groundwater. It cannot control fungal species orthe insects but its use as nematicide is appreciated by many. These formulations can be in the form of liquid and they exist as 2 isomers. In the vapour phase, the trans form is more toxic to the nematodes and the toxicity can be about 90%. Ethylene dibromide is available in all parts of the world and their use as nematicides is still in progress. They can contaminate groundwater and can be carcinogenic to h1,2-dibromo-3-chloropropane can be used in the form of liquid formulations. They can be are used in the form of fumigants. They cantack the nematodes. Though they are effective against the neodes, their activity onthe humans is questionable as it causes sted cancers inthe human beings. Methyl bromide is another fumigant which canbe toxicin nature and can kill several nematodes. We can use it to combat the menace caused by the weeds, insects, nematodes and soil fungi. This chemical can cause damage to the ozone layer and hence banned according toeal protocol. This chemical is being used as gas. They don't have a specific odor. In order tTo identify the presence of this chemical, chloropicrin is being used with methyl bromide. This chemical can move freely in the soil and at a faster pace.

Chloropicrin is considered as the fumigant since ages. It can be used as a fungicide. There is some activity on the nematodes and weeds. We can use this chemical in the 1,3 D formulations so that there will be enhanced fungal activity. This compound can be toxic and can be sprayed in the form of liquid formulations. Methyl isocyanate, Dazomet and Metam sodium are the three chemicals which can impose activity on the nematodes and destroy them. The chemical metam sodium can be classified as a soil fumigant. It can be used to regulate the weeds, insects, fungi and several other nematodes. The metam sodium can be converted into methyl isocyanate. This chemical can be considered as a biocide. This chemical is being used as soil pesticide. The compounds which are the same as the metam sodium can also inflict damage to the nematodes. Dazomet can be used as a fumigant and it can be applied in the form of granules. Some of the chemicals such as glucosinolates and isothiocyanates can be used as nematicides. The plants which belong to the brassicaceous family can produce thepounds when linked w the soils.

Sodium tetrathiocarbonate can be considered as a fumigant which that can be applied on to the soils. It can kill nematodes, insects and fungi. They can be used as liquid formulations. We can use these chemicals when we practice surface irrigation or drip tyrigation. This chemical can get degrade degradation, there will be thduction of sulphur sulfurand other gases such as hydrogen sulphidessulfides. The disulfidecane can used as a fumigant and it is more effective but the main problem is that it can getcatches fire. Ald can be used as insecticide and nematicide. It can kill both mites and insects. It is a chemical which that can be used all over the world to control nematodes. We can use this chemical in the form of granules. They exhibit systemic activity also. Oxamyl, carbofuran and aldicard are considered as promiematicides than others as their activity is instant and effectxycarb is another chemical which that can be used as a nematicide. The chemical aldicarb can be oxidised oxidized in the soils. Due to the oxidation reactions, there will be the formation of aldicarb sulfones.We can use it in the form oe formulations. Carbofuran can be classified as nematicide and insecticide. They can be systemicin nature. They can be used in the form of liquid formulations and granular formulations. They are toxic to the bird species. Oxamyl can be classified as a carbamate. We can use this chemical in the form of liquid formulations and granular formulations. It can exhibit systemic activity. We can use this chemical as a foliar nematicide. It can be used anywhere in the world. Ethoprop can work as both nematicide and insecticide. It may not exhibit systemic nature. The half- life of this chemical is based on the organic matter present in the soils. We can use this chemical in the form of emulsifiable concentrates and granules. Fenamiphos can be considered as insecticide and nematicide. It can be stored on the organic matter which is coas important and beneficial step compared to others. It can be toxic to humans but not carcinogenicin nature. Cadusafos can be considered as an organophosphate. It doesn't exhibit systemic se this chemical to kill the insects and nematodes which live in the soils. This chemical has ocannot contaminate groundwater and it can regulate the nematodes which attack the citrus crop. We can use Tthis nematicideis usein the form of microencapsulated formulations and granular formulations.

Fosthiazate can be classified as an organophosphorus nematicide. It can exhibit broad- spectrum activity. We can use this chemical as microgranule formulations. It ca be used to control the lesions caused by the nematodes. The nematode which can be killed by this chemical is pPratylenchuspenetrans. This nematode can act on the potatoes and reduce the yields dramatically. The micrograe formulations of this nematode can be clay- based. This nematicide cannot act on the nematodes which attack on the tobacco plants. Terbufos can be classified as organophosphate nematicide. It can only work as an insecticide. It can be used as emulsifiablentrates and granular formulatios. We can use this chemical as a soil insecticide. DiTera is a nematicide which that can control nematode nam The toxic nature of this chemical be useful to initiate synergistic action on the compounds which are lighterlighter compounds. It cannot act onthe mammalian and free- living nematodes. We can observe some of the toxic effects in G.rostochinensis. The movement of this species wilbe impaired due to this chemical. We can use this nematicide in the form of emulsifiable suspensions, powders andas granules. Clandosan can be used as a nematicide. They can be used as a granular product. It is produced from the crawfish exoskeletons and crabs. This chemical is derived from these exoskeletons is dominated by urea and chitin. This chemical can exhibit nematicidal activity. This ch mustcarefully applied so that there will not be any sort of photoactivity. Sinocin is another nematicide which that can is produced from the extracts derived rom Rhizophora mangle, Rhus aromatic and Opuntia lindheimeri. This chemica can regulate the nematodes which attacks the citrus plants. The developmental stages of the nematodes can be proned to the nematicides.

The fumigants can be considered effective against the nematodving in the soils. The halogenated hydrocarbons can be used as alkylating agents and thich that can regulate the soil nematodes. The organophosphates and carbamates can work as inhibitors which are reversiblereversible inhibitors. They clower the acetylcholinesterase activity mainly in the insects. The nematicides which are classified as non-fumigant can be inhibit these neurotransmitters such as cholinesterase. Some of the chemicals which can be used as non-fumigants are oxamyl, fenamiphos, carbofuran and aldicarb. Carbofuran compounds can inhibit the neurotransmitters in the Meloidogyne species. Some of the symptoms which can be seen post nervous damage in the insects are uncoiling, soiling, convulsions, trembling, twitching and stylet thrusting. The concentration of the contact nematicide in the soils can kill the nematodes. Soil can be considered as a mixture which comprises of a heterogeneous mixture. The nematode population can be completely controlled by the nematicides. The contact nematicides may not inflict complete damage. The contact nematicides can restrict the penetration and movement of the nematodes and these actions are enough to control the growth of nematodes. The fumigation of the soil needs perfect preparation to be effective. The soil must be tilled so that will be an increment in the porosity. There will be uniformity established. The nematodes can be classified as invertebrates which that are present in both aquatic and terrestrial habitats. The nematode can dominate the soils than any other species. They comprise of simple body exterior and interior so that they can absorb the chemicals or any dissolvedials in the capillary soil water. They can feed on the plants, algae, animals, fungi and bacteria. There are about 4100 species which that can attack plants. They can cause both direct and indirect damage to the crops. Some species can infect the stems and le mostly these species can attack the rhizosphere region. Due toamage caused by these nematodes, there willbe loss will hich can be about million dollars each year. Th several meich that can be used to regulate the plant attacking nematodes. We can classify these methods into biologicalmethods, cultural methods and chemical control methods.

The soil fumigants can be used to lower the nematode populations. Some plants can produce organic molecules which are classified as secondary metabolites. These molecules can be used for the development of plants but not essential for the survival of the plants. The metabolism of the plants at its initial stages can regulated by these secondary metabolites. These molecules can be usefulr the productioncides. These compounds can protectthe plants from several diseases and pests. These chemicals can be crucial in to maintain the defences defensesbetween the species. Some of the secondary metabolites are polyphenols, diterpenoids, flavonoids, monoterpenoids and alkaloids. Tannins can be useful for plants in many ways. These chemicals can kill yeasts, bacteria and fungi. Tannins can control the growth of the nematodes. The nematicidal effects can be inflicted by glucosinolates, flavonoids, diterpenes and steroids. The secondary metabolites can influence the nematodes by changing thei patterns. They can act as repellnd attractants. Due to these actions, there will be damage to the nematodes and render them paralysedparalyzed. There are some plants which that can be used to lower the damage incurred due to the nematodes. These plants can control the nematode population in several crops. Marigolds, castor bean and chrysanthemum and other species such as crotalaria can be used to control the growth of the nematodes. They can produce some compounds which can be toxic to the nematodes. These compounds can be produced by the roots. We can use these plants as green manure. These plants may not be edible for the nematodes. Some of the bacterial species can interrupt the growth of the nematodes by interfering with their life cycles. Marigold is considered as an important plant species which can control the growth of the nematodes. Marigold can suppress the garden nematodes likedogyne and Pratylenchus. These plant species can also control stunt nematodes. This species can be used as trap crops. They release chemicals which that are toxic to the nematodes. All the marigold plants cannot reduce the nematode populations. Marigold can be susceptible to the nematode species such as Rotylenchus, Trichodorus and Criconemella. Tagetespatula can control most of the nematodes. Some of the French dwarf marigolds can reduce the growth of these nematodes.

Some of the marigold species such as T.minuta and T.erecta can lower the nematode populations. Some of the tropical legumes can be used to improve the fertility of the soils. They can lower the damages inflicted due to nematodes. The leaves erated from these legumes can serve as green manure. They can lr the populations of some of the nematodes. There are many legumes which can grow inthe tropical conditions. Some of these plants can be vulnerable tothe nematode infections. The susceptible species are alyceclover, pigeonpea, hairy indigo and carpondesmodium. Castor beans can be used to lower the nematode populations. These plancan reduce the lesion nematodes and peanut nematodes. These beans can be placedunder the space provided by the green manure. Each seed can be toxic tothe livestock and pests. The peanut root-knot nematode can be killed by AeschynomemeAamericana. Cassia fasciculateor partridge pea can be used to kill the peanut root-knot nematode. C.spectabilis can be susceptibto root-knot nematodes. Some of the root-knot nematodes which can kill this species are southern root-knot and Javanese rot-knot. Mucunadeeringiana can be control the nematode populations with the help of rhizobacteria. It also reduces the root-knot nematodes which can kill C. Sspectabilis. The temperate legumes can be used as a cover crop in the winter. Common vetch can be resistant to several root-knot nematodes. Some of the root-knot nematodes which can be killed on treating with common vetch are Javanese, peanut and southern root-knot nematodes. This plant can be treated as weeds. Hairy vetch can be damaged due to the activity of the nematodes. It is important to know that these species should not be grown in the areas already infected with the nematodes. Rapeseed or Brassica napus can be grown as an annual crop.

5.3 EVALUATION

The efficacy of the nematicides can vary according to the chemical composition. Chemicals such as hydrogen peroxide, chitosan, boran and abamectin are considered as important nematicides. These chemicals can control the growth of the citrus nematode which is called Tylenchulussemipenetrans. Some of the species which can infect the citrus species are Meloidogyne spp., Pratylenchus coffee and Radopholussimilis.T. semipenetrans can be incurring losses in the citrus crops. The management of this species requires many efforts. It is easy to go for alternative methods to control these nematodes. Abamection is a nematicide produced from bacteria Streptomyces avermitilis. This nematicide can control the growth of several phytonematodesThis chemical can be used to treat the seeds. It is a gopractice to use this chemical as a soil fumigant. Boron is classified as a mnutrient. It is crucial inthe plant metabolism. This chemical is important to enhancethe crop yields. With the use of boric acid, the root galls in the grape crops was were reduced drastically. The derivatives of chitin such as oligochitosan, chitosan can be biocompatible and biodegradable. The molecule chitosan can exhibit nematicidal and nematistatic action. This chemical can be used to manage diseases. It can act on the nematodes which act on the plants. Hydrogen peroxide can be used to regulate the plant nematodes. 2000o ppm of hydrogen peroxide can be useful to comb growth of the nematodes. This chemical can enhance the systemic resistance in the sugar beet plants. The root-knot nematode can be considered as an important pest which that affects many crops in the southern U.S. These nematodes can inflict losses in the soybean, corn and cotton crops. This nematode can be classified as endoparasite. It can feed and reproduce in the plant cells ch are previously altered in the roots of the plant. This nematode can prodots and galls. It can infect about 3000 species. Nematicides can be crucial in regulatingthe plant-parasite relations. It is a good practice to use volatile toxicants which that are broad -spectrum and can be mixed within the soin be non-selective and most of the nematodes die. These nematodes can be active even after 30 days of application. The nematicides can be used in four forms which are commoncommon forms. The important forms are fumigants, granules, biological and seed treatments. The nematicides whiche being used in the granular form can be useo protect both nematodes and rootworms. The treatment of seed has become common to combat the nematodes due to enhanced usage ofthe pesticides. These nematicides can be safer to for the environment. The fertilizers can be to treat the seeds so that the nutrients will be available to them. When the nutrients are applied to the seeds, the seedling when generates can use the nutrients. Because of the ty of the nutrients, the crops can grow easily and hence there will be a competitive advantage over weeds. We can use the starter fertilizers to enhance the grain yields. In order tTo increase the grain yields, it is crucial that optimal temperatures are maintained. If the soils are having lower temperatures, then the organic nitrogen will be converted into inorganic nitrogen. The phosphates which are in the solid state will be converted into a liquid state.

The starter fertilizers can offer the crops several pathological advantages. The starter fertilizers can control the nematode populations in the corn species. The root-knot nematode can induce galls and so the lateral roots arising from the roots will be reduced. Due to the formation of the galls, there will be an imbalance of nutrients. We can observe metabolic sink in these plants. The starter fertilizers can help the host plants so that the side effects can be minimized. The plant auxins or growth regulators can play a crucial role in combating nematodes. The growth of the plants can be regulated by the cytokinins, gibberellic acid and auxins.

6

Rodenticides

CONTENTS

6.1 INTRODUCTION

The rodenticides can be classified into anticoagulants, metal phosphates, hypercalcemia and others. There are rodenticides which can exhibit low rate of toxicity. The anticoagulantsThe rodenticides can be classified into anticoagulants, metal phosphates, hypercalcemia and others. There are rodenticides which can exhibit low rate of toxicity. The anticoagulants can be chronic or acute. In the case of chronic anticoagulants, death due to the rodenticide may occur in two weeks. In the case of acute anticoagulants, rodents can die immediately. They can be used as single or multiple doses.e single single-dose rodents can be called as second -generation rodenticides. The multiple -dose rodenticides can be called as first -generation rodenticides. They can block the vitamin K cycle. Because of the blockage of the vitamin K cycle, there will be loss of the blood clotting factors and hence rodents can die because of blood leakage. There are two important factors needed in the coagulation process. They are prothrombin and proconvertin. Prothrombin can be called as factor II and proconvertin can be called as factor VII. These rodenticides also affectthe humans, cats and dogs. This aspect can be called as secondary poisoning risk. When these rodents are killed, they will be eaten by secondary consumers which will be affected in the same way as the rodents.

Figure 6.1 Rodent- Rattus norvegicus.

Source: Source:http://saferodentcontrol.org/site/rodent-control/

The blood capillaries will be disrupted due to indandione, thiochromenone and hydroxycoumarin chemicals. These chemis can enhancethe permeability so there will be severe internal bleeding. These harmful effects are slow and can take some time to kill the rodents. The rodentic can suffer with from severe anemia and hemorrhagic shock. The advantage of the anticoagulants is that the poison can impose death. The rodenticidal anticoagulants can exhibit minimal half- life. The second- generation rodenticides can exhibit lower toxicity. The first generation rodenticides can be less toxic than the second generation. Superwarfarins are considered as second -generation anticoagulants. Coumlyl and warfarin can beare considered as first -generation rodenticides.

Type of rodenticide	Active ingredient	Formulations
First-generation anticoagulant	Warfarin	Loose grain
	Coumatetralyl	Loose grain, Contact powder
	Chlorophacinone	Liquid concentrate
Second-generation anticoagulant	Difenacoum	Loose grain, Pellets, Wax blocks, Place packs
	Bromadiolone	Loose grain, Contact powder, Pellets, Wax blocks, Place packs
	Brodifacoum	Loose grain, Pellets, Wax blocks, Place packs
	Flocoumafen	Wax blocks
Non-anticoagulant	Calciferol	Liquid concentrate, Pellets
Fumigant	Aluminium phosphide	Tablets

Figure 6.2. Active ingredients.

Source: Source:http://saferodentcontrol.org/site/rodent-control/

The chemicals such as bromadiolone, flocoumafen, brodifacoum and difenacoum can be considered as second -generation rodenticides. Indandiones can be classified as second -generation rodenticides. Difethialone can be classified as 4-thiochromenones. Sulfaquinoxaline can be classified as a bacteriostatic or antibiotic agent. This chemical can disrupt the intestinal microflora and these organisms are important for the production of vitamin K. Because of the reduction of the microbial activity in the intestines, the anticoagulant property will get increased. This activity can be enhanced by vitamin D. This vitamin can exhibit a synergistic effect.

Vitamin K_1 was u control several pests which that attack humans and these poisons can ave anticoagulants. These posisons can lower the activity of the liver. Metal phosphides can be used to kill the rodents. These chemicals can be classified as single- dose rodenticides. They can act faster. Death can occur after three days. Generally, we can use bait tagged with food and metal phosphide. Mostly zinc phosphide can be used. When rodents ingest these phosphides, the acid present in their stomach reach with the phosphides and generate phosphine gas. The baits made up of zinc phosphide are cheaper. It can be usefulr large scale elimination ofthe rodents. We can add atleast 0.75% concentration of zinc phosphide in the bait. These baits can emit garlic odor which is pungent and this is due to the hydrolysis rtions. This odor can attractthe rodents. There are pellets and tablets available which can produce ammonia. It is considered safe to upellets or tablets as phosgene phosphine gas emitted from aforesaid chemicals can be dangerous. These metal phosphides cannot get accumulated in the animal tissues and so if secondary consumers eat the bodies of these rodents, they may not be affected. As strychnine trees were damaged by the Japanese, the United States started using phosphides. These chemicals can be classified as fast -acting chemicals. Some of the phosphides which can be classfied as rodenticides are aluminium phosphide, calcium phosphide, magnesium phosphide and zinc phosphide. The first three can be used as fumigants and the last one must be used as bait.

Figure 6.3 Pellet consumed by rat.

Source: http://saferodentcontrol.org/site/rodent-control/

Hypercalcemia can be effective onthe rodents. Vitamin D_2 and D_3 can be classified as denticides. Vitamin D_2 is calledas ergocalcerol and vitamin D_3 can be calledas cecalciferol. They can be toxic tothe rodents. The homeostasis of the rodents can be affected by these chemicals. Hypervitaminosis can occur if vitamin D is used in higher concentrations. When rodents ingest these chemicals, there will be hypercalcemia. Due to this condition, there will be an increase in the levels of calcium.

Figure 6.4. Bait boxes.

Source: http://saferodentcontrol.org/site/rodent-control/

Figure 6.5. Bait Containers.

Source: http://saferodentcontrol.org/site/rodent-control/

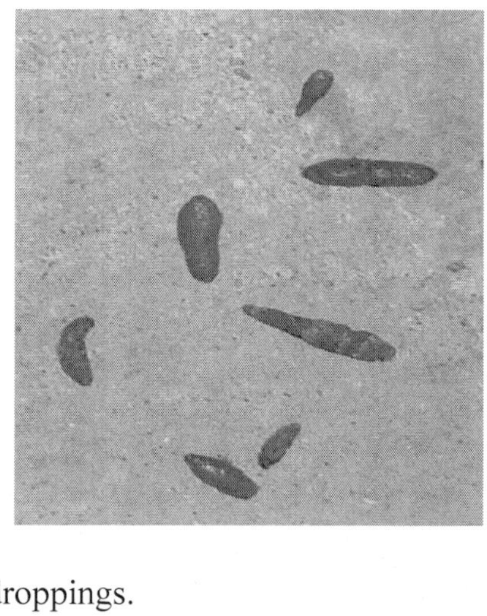

Figure 6.6. Colored droppings.

Source: http://saferodentcontrol.org/site/rodent-control/

The calcium content present in the food materials can be taken up by the rodents. The calcium content present in the bone matrix can be transformed into the ionized form. These ions will be allowed to get circulated in the plasma. The enhanced calcium levels can get deposited in the lungs, stomach wall, kidneys and blood vessels. These organs and organ systems will get calcified and mineralized. There will be the formation of calcium crystals and calcificates. This results in severe heart problems,kidney failure, bleeding and damage to myocardial tissues. We can use 0.1% ergocalciferol and 0.075% cholecalciferol to kill these rodentshese anticoagulants can be toxic tothe pets and humans. The baits which comprise of caerol can be considered as safe for avesbirds. Some of the other chemicals which can be used as rodenticides are thallium sulfate,Bromethalin,rat ice, strychnine, sodium fluoroacetate and pyrinuron. There are several other chemicals which can be considered as rodenticides.We can use these chemicals in various combinations. Vitamin D+ antibiotic+ anticoagulants can be used in combination. These combinations can act faster on the rodents. Some of the rodenticides which can exhibit a low rate of toxicity are corn meal gluten and powdered corn cob. These chemicals can create an imbalancf the electrolytes and death. Because ofthe secondary poisoning, the wildlife will be at risk.

6.2 CLASSIFICATION

The rodenticides can be classified into botanical rodenticides, carbanilate rodenticides, coumarin rodenticides, indandione rodenticides, inorganic rodenticides, organochlorine rodenticides, organofluorine rodenticides, organophosphorus rodenticides, pyrimidinamine rodenticides, thiourea rodenticides and urea rodenticides.

Rodenticide	Type	Chemical Class	Days of feeding needed
Warfarin	Anticoagulant	Hydroxycoumarin	multiple
Chlorophacinone	Anticoagulant	Indandione	multiple
Diphacinone	Anticoagulant	Indandione	multiple
Bromadiolone	Anticoagulant	Hydroxycoumarin	single
Difethialone	Anticoagulant	Hydroxycoumarin	single
Brodifacoum	Anticoagulant	Hydroxycoumarin	single
Bromethalin	Non-anticoagulant	other	single
Cholecalciferol	Non-anticoagulant	Vitamin D3	multiple
Zinc phosphide	Non-anticoagulant	other	single
Strychnine	Non-anticoagulant	other	single

Figure Figure 6.7. Toxicity of rodenticides.

Source: http://npic.orst.edu/factsheets/rodenticides.html

	Oral	Inhalation	Dermal	Primary Eye Irritation	Primary Skin Irritation
Warfarin[3,6]	Moderate - High toxicity	Not significant	Not significant	No data	No data
Chlorphacinone[4]	High toxicity	High toxicity	High toxicity	Non-irritating	Non-irritating
Diphacinone[4]	High toxicity	High toxicity	High toxicity	Moderate irritation	Slight irritation
Bromadiolone[4]	High toxicity	High toxicity	High toxicity	Low irritation	Minimally irritating
Difethialone[3]	High toxicity	High toxicity	High toxicity	Mild irritant	Non-irritating
Brodifacoum[4]	High toxicity	High toxicity	High toxicity	Minor irritation	Mild irritant
Bromethalin[4]	High toxicity	High toxicity	Moderate toxicity	Slight irritation	Non-irritating
Cholecalciferol[3]	High toxicity	Very low toxicity	Low toxicity	No data	No data
Zinc phosphide[5]	High toxicity	High toxicity	Low toxicity	Slight irritation	Non-irritating
Strychnine[7]	High toxicity	High toxicity	Low toxicity	Highly irritating	Non-irritating

Figure 6.8. Common rodenticides.

Source: http://npic.orst.edu/factsheets/rodenticides.html

Rodenticide	Secondary risk to birds	Secondary risk to mammals
Warfarin	low risk	low risk
Chlorophacinone	low risk	high risk
Diphacinone	moderate risk	high risk
Bromadiolone	moderate risk	high risk
Difethialone	high risk	high risk
Brodifacoum	high risk	high risk
Bromethalin	low risk (based on limited evidence)	low risk (based on limited evidence)
Cholecalciferol	low risk (based on limited evidence)	low risk (based on limited evidence)
Zinc phosphide	low risk	low risk
Strychnine[17,24]	possible (insufficient data)	possible (insufficient data)

Figure 6.9. Secondary poisoning.

Source: http://npic.orst.edu/factsheets/rodenticides.html

There are several unclassified rodenticides. Botanical rodenticides include scilliroside and strychnine. The carbanilate rodenticides include mieshuan. Coumarin rodenticides include brodifacoum, bromadiolone, alpha bromadiolone, coumachlor, coumafuryl, Coumatetralyl, documarol, difenacoum, difethialone, flocoumafen and warfarin. Indandione rodenticides comprise of chlorophacinone, diphacinone, naphthylindanediones, pindones and valone.

Inorganic rodenticidesan arsenous oxide, phosphorus, potassium arsenite, sodium srsenitearsenite, thallium sulfate and zinc phosphide. The organochlorine rodenticides include lindane, HCH and gamma HCH.

Some of the organofluorine rodenticides are sodium fluoroacetate gliftor, fluoroacetic acid and fluoroacetamide. The organophosphorous rodenticides comprise of phosacetim. Pyrimidinamine rodenticides include crimidine. Thiourea rodenticides include thiosemicarbazide, promurit and antu. Urea rodenticides include pyrinuron. Some of the unclassified rodenticides include bromethalin, chloralose, cholecalciferol, curcumenol, ergocalciferol, flupropadine, hydrogen cyanide, tetramine, silatrane, paris green and norbormide.

Rodenticide	Classification	Mode of action	Acute oral toxicity	Antidote	Primary poisoning risk	Secondary poisoning risk
Warfarin			Moderate to high		Low (birds), moderate (mammals). Highly toxic to cats	Moderate (birds and mammals)
Chlorophacinone ([1]Restricted Use Pesticide)			High		Low (birds and mammals)	Low (birds), high (mammals)
Diphacinone ([1]Restricted Use Pesticide)			High		Low (birds and mammals)	Moderate (birds), high (mammals)
Brodifacoum	Anticoagulant	Inhibition of blood clotting and capillary damage	High	Vitamin K	High (birds and mammals)	High (birds and mammals)
Bromadiolone			High		Moderate (birds), high (mammals)	Moderate (birds and mammals)
Difenacoum			High		Moderate (birds), high (mammals)	Moderate (birds), unknown (mammals)
Difethialone			High		High (birds), moderate (mammals)	High (birds), moderate (mammals)
Strychnine		Convulsant and respiratory failure	High	None. Supportive treatment for symptoms.	High (birds and mammals)	High (birds and mammals)
Bromethalin		Respiratory arrest	High		Low (birds and mammals)	Low (birds and mammals)
Cholecalciferol	Non-anticoagulant	Hypercalcemia	High	Calcitonin	Low to moderate (birds and mammals)	Low (birds and mammals)
Zinc phosphide ([1]Restricted Use Pesticide)		Prevents cellular energy production	High	None. Supportive treatment for symptoms.	High (birds and mammals)	Low (birds and mammals)

Figure 6.10. Rodenticides with modes of action.

Source: https://edis.ifas.ufl.edu/pi284

Species (body weight, kg)	Zinc phosphide (2.5%)	Diphacinone (0.005%)	Warfarin (0.025%)	Brodifacoum (0.005%)
Rat (0.25)	0.45	16–25.5	58	1.4
Mouse (0.025)	A	70	37	0.2–0.43
Rabbit (1)	A	700	3,200	5.8
Pig (50)	40–80	150,000	200–1,000	500–2,000
Dog (5)	4–8	88	400–5,000	25–100
Cat (2)	1.6–3.2	588	48–320	1,000
Chicken (1)	0.8–1.2	A	4,000	200–2,000

Figure 6.11. Rodenticides –safe LD_{50} values.

Sorce: Source:https://edis.ifas.ufl.edu/pi284

Mammalian toxicology	LD50 in mg/kg b.w.		
oral	Difenacoum	Bromadiolone	Brodifacoum
Rat	1.80	1.125	0.27
Mouse	0.80	1.75	0.40
Rabbit	2	1	0.30
Dog	>50	>10	0.25-1.00
Cat	100	>25	25
Guinea pigs	50	/	2.8
Dermal			
Rabbit	1000	2.1	/
Rat	27.4	/	0.25-0.63

Figure 6.12. LD_{50} comparision.

*Source: https://www.arcjournals.org/pdfs/ijrsz/v2-i3/1.***.3**

6.3 MODES OF ACTION

The coumarin anticoagulants were developed in theSecondWorld WarII. They were used to treat thromboembolic disease. These chemicals can work as antithrombotic agents. Warfarin was used as rodenticide and drugs. Indandione and hydroxycoumarin derivatives were produced long ago and they act as powerful rodenticides. The blood coagulation system will be damaged on exposure to these chemicals.

There was a demand for throduction of superwarfarins or anticoagulantsThese chemicals were used inthe agricultural practices and rodent control inthe urban areas. These rodenticides were classified into two main classes.

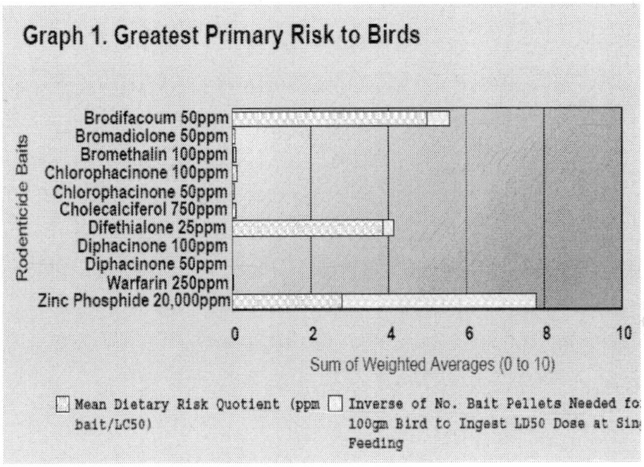

Figure Figure 6.13. Risk to birds.

Source:https://www.fwspubs.org/doi/suppl/10.3996/052012-JFWM-042/suppl_file/10.3996_052012-jfwm-042.s4.pdf

They are hydroxycoumarins and indandiones. These two classes share almost similar mechanisms. These rodenticides can be used in various formulations. We can use them as powder form and solid crystalline form. We can determine these rodenticides with the h HPLC or high performance liquid chromatography.

The baits have ingredients which that are active and they determine the strength of the chemicals. These chemicals can be used in the form of bait formulations. They exhibit a low rate of volatility and so t overall air concentration will be lower. They exhibit no solubility in water andin some cases are least soluble. We can expect a low rate of water contamination.

No crops the rodents and foodstuffs will be contamination free. This is due to the fact thatbecause these chemicals will be sprayed when cropping is not present. Some of the vertebrates will be affected by these rodenticides as they ingest bait by accident. We can use these chemicals in the whole grain bait and as small pellets. It is important to know that there will be occupational hazards due to these rodenticides.

Most of the rodenticides can be classified as vitamin K antagonists. They will act on the liver. Most of the precursors for blood coagulation are located in the liver. There will be post- translation processing which needs vitamin K. After the processing occurs, there will be the formation of procoagulant zymogens.The K_1 epoxide reductase will be inhibited due to these rodenticides. The gastrointestinal tract can absorb these rodenticides. The respiratory systemand skin can easily absorb the rodenticides. These chemicals must be eliminated through faeces normally. In the case of rats, there will be hydroxylation of the indandiones and warfarin as part of metabolism. These rodenticides can remn unchanged and can be excreted. The rodenticides can get accumulated in the fat tissues andin the liver.

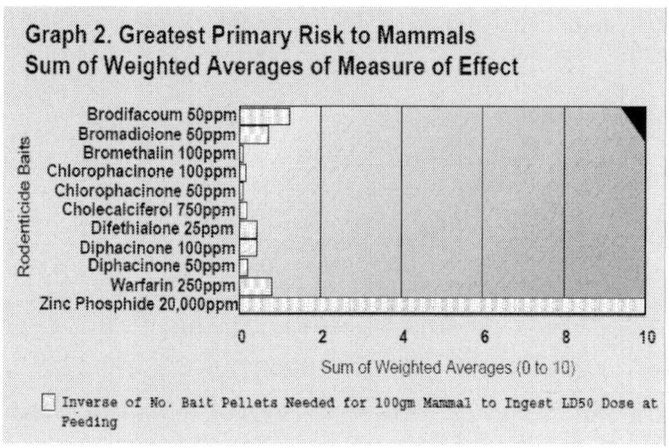

Figure 6.14. Risk to Mammals.

Source: https://www.fwspubs.org/doi/suppl/10.3996/052012-JFWM-042/suppl_file/10.3996_052012-jfwm-042.s4.pdf

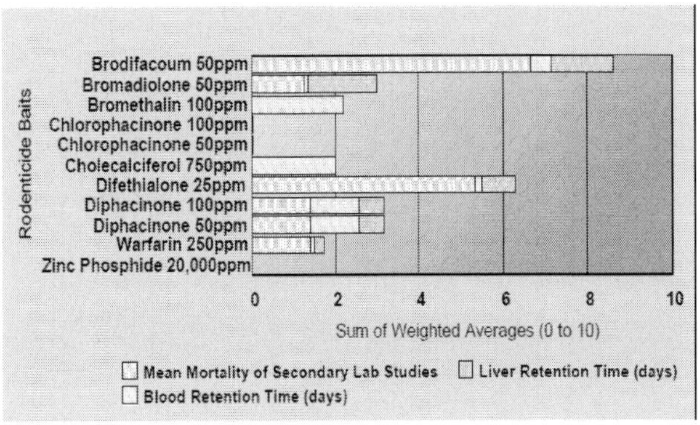

Figure 6.15. Secondary risk to birds

Source: https://www.fwspubs.rg/doi/suppl/10.3996/052012-JFWM-042/suppl_file/10.3996_052012-jfwm-042.s4.pdf

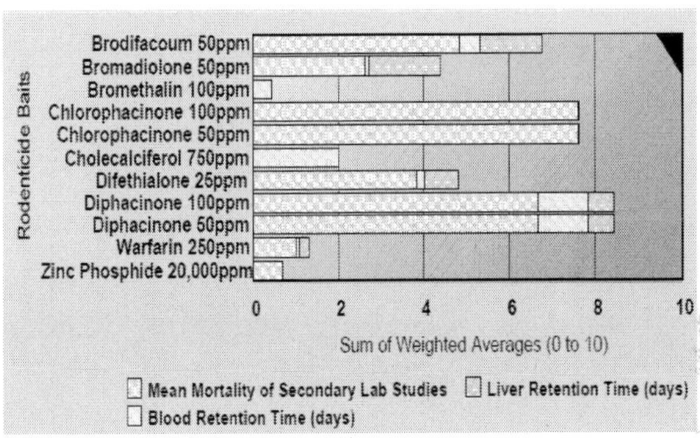

Figure 6.16. SSecondary risks to Mammals.

Source: https://www.fwspubs.org/doi/suppl/10.3996/052012-JFWM-042/suppl_file/10.3996_052012-jfwm-042.s4.pdf

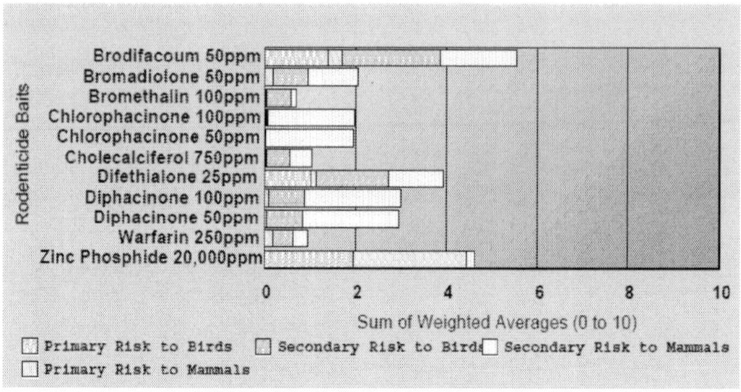

Figure Figure 6.17. Totalrisk to Birds and Mammals.

Source: https://www.fwspubs.org/doi/suppl/10.3996/052012-JFWM-042/suppl_file/10.3996_052012-jfwm-042.s4.pdf

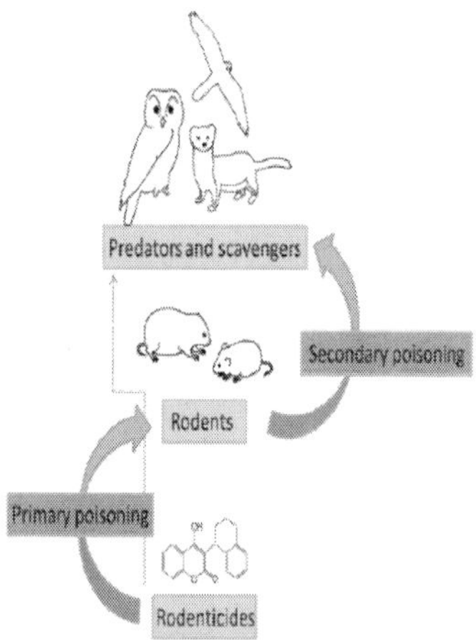

Figure 6.18. Routes of exposure

On exposure to these chemicals, rats will suffer from intense bleeding. The rate of toxicity is higher when taken orally. The toxicity rate is also higher when these rodenticides are inhaled. The rate of toxicity can be based on the diet, sex and strain of the rodents. Waarin can be present in the bloodstream for long periods. In the case of acute poisoning, we needto vitamin K_1 to neutralize the effects. Several incidents of rodenticides poisoning were reporte. Some of them are intentional. The main symptoms of these rodenticides poisoning can be massive haemorrhage and intense bleeding. These symptoms can get developed after 3 days of exposure. The effects of these rodenticides can be classified into primary effects and secondary effects. The primary effect is due to direct poisoning and the secondary effect is due to the intake of the rodents killed by the rodenticides. The rodenticides can be divided into hydroxycoumarins and indandiones as mentioned earlier. These rodenticides can be supplied in the form of solids. The solid forms can be classified into powders and crystals. These chemicals can be sparsely soluble in water. They can be dissolved in acetone readily. Warfarin can be classified as an acid. It is insoluble in distilled water. When pH is higher, this chemical can be ionized and hence may dissolve in water. Some of the contaminants such as detergents and proteins can facilitate the dissolution of Warfarin in water.

Most of the rodenticides may not be available naturally. Based on the toxicity rate, the active ingredients can be atleast 0.005% in the case of indandiones. The rodenticides can be used in diverse formulations. They can be used in the form of tracking powder, pelleted baits, whole grain baits and paraffin wax blocks. Some companies use bittering agents to the bait. Bitrex or denatonium benzoate can be used. The whole baits and small pellets can attract several birds. These rodenticides can also attack other vertebrates which that are not targets. We can use wax blocks instead of the pellets and baits so that birds won't be attracted. The wild rodents and commensal rodents can lead to the death of cats and birds. The dry baits can be used to she grain stores, rarely; there may be an occurrence of high concentrations. The staff working in the industriesindustries can be exposed to bait appland formulations.

Vitamin K can be used for various compounds. These compounds can work as co-enzymes. Crucial The crucial enzyme required inhis process is the gamma glutamate carboxylase. These chemicals can have a ring structure which can be calledas naphthoquinone. Vitamin K_1 can be called as phytomenadione. In this molecule, there will be 4 isoprenoid residues. Ouof these 4 isoprenoid residues, one residue can be in an unsaturated state. Vitamin K_2 compounds can be calledas menaquinones. There are 13 isoprenoid residues and mostly they are unsaturated. Vitamin K_3 can be called as menadionehis molecule cannot have side chains. The indandiones and 4-hydroxycoumarin derivatives can be cassified as vitVit-K antagonists. The important step exhibited by these chemicals is that they can inhibit the vVit-K dependent reactions. Some of the factors which are involved in the coagulation cascade are coagulation inhibiting proteins S and C, X or the stuart-prower factor, IX or the Chfactor, VII or proconvertin and II or prothrombin. These proteins can be produced in the liver tissues. Prior toBefore their release into the bloodstream, these precursor proteins can go for post -translational modifications. In all these cases vitamin-K can acues are important for this entire activity. KH_2 or vitamin K hydroquinone can work as acoenzyme which is activn active coenzyme. This co-enzyme can get oxidized and the energy generated can be used in the carboxylation reactions. There will be the formation of an epoxide. This is a two- step process and it is mediated by KO reductase. Due to the blockade of this enzyme, there will be excess utility or exhaustion of the KH_2 supply. These reactions will stop the production of the GLA residues. There will be deposition of the coagulation factors which are not carboxylated. These chemicals will appear in theood circulation. In this case, we call these factors as descarboxy coagulation factors. These factors can be inthe circulation in the form of zymogens. The zymogens can be activated only if they undergo certain proteolytic degradation reactions. The coagulation factors which ae considered as descar can exhibit no procoagulant activity. They may not be converted into zymogens. Leck and pPark in the year 1981 has have explained fects of blood coagulation factors and k metabolism in the susceptible and resistant rats. In the case of rats which that are susceptible, there was an increase in the levels of warfarin brodifacoum. Due to this activity, there was enhanced KO circulation. This explains that these drugs can influencethe KO reductase activity. The KO and K ratio is higher in the rats which are susceptible. Park and Leck in 1982 also reported that the chemicals brodifacoum and difenacoum can act as vitamin K antagonists. The second -generation anticoagulants can exhibit strong lipophilic nature. Due to this nature, they can firmly bind to the membranes. KO reductase can be classified as an integral membrane protein.

The rodenticides can be absorbed via the respiratory system, skin and gastrointestinal tract. Warfarin can be hydroxylated in vivo and in vitro by the liver. Due to the enzymes present in the microsomes, there will be the formation of 6, 8 and 7 hydroxy arfarin. These metabolites in their inactive state can bond with glucuronic acid. These chemicals will be excreted as faeces and urine. Indonesand diones can be metabolized due to the hydroxylation process. Brodifacoum can be observed in the sheep's liver even after 128 days. It was observed that the second -generation anticoagulants can be toxic than the first eneration coagulants. Some of the symptoms of this poisoning are vein clearing, hunched posture and lethargy. There will bbe blood around the anus, mouth and eyes. The inhalation of the second- generation anticoagulants was proved more toxic and acute.

Broifacoum can act as a skin iritant and eye irritant. Warfarin can enhance the rates of structural malformations, internal haemorrhage and subcutaneous haemorrhage and embryolethality. Injecting warfarin or brodifacoum post Phenobarbital treatment can exhibit reduced anticoagulant effects. Some of thedrugs which are considered as anti- inflammatory are phenylbutazone uprofen. These drugs can be considered as lethal for the rodents. Sex and strain can be considered as crucial factors which that can determine the toxicity rates in rodents. Vitamin K_3 can be added in massive amounts to the feed. When accidental exposure occurs to the animals, post-tion, this can serve as an antidote. Due to this activity, there will be reduced lethality. There are several factors which that can be considered whilst explaining inhibition of the production of vitamin K. Some of the factors include faster rodenticide clearance, reverse inhibition of epoxide hydrolase and lowered sensitivity of vitamin K epoxide hydrolase.

The primary cause of the poisoning of the non-targeted species is the consumption of the baits which are loaded with the anticoagulants. The anti-coagulant rodenticides can cause bleeding for a long period. It is important to treat the cattle for several weeks so that the chemicals in the body get eliminated. Some of the symptoms which can appear after 12 hours are anorexia, weakness, cold gums, pale mouth, labored breathing, excessive bleeding. Blood in the urine and stools and bruising. Phytomenavitamin K_1 can be considered as an important antidote. It is safer than the other vitamins. Mount and Feldman in 1983 has have have explained that the therapy which is effective on the accidental warfarin poisoning may not be effective on the accidental diphacinone poisoning. This can be due to thedifference in the therapy duration and vitamin K_1 dose. In Egypt, there were several abortions reported in the goats and sheeps due to accidental brodiacoum poisoning. Some of the symptoms associated were abortion, abdominal bloating, lameness, flank ecchymosis, subcutaneous haemorrhage, recumbency, haematochezia and epistaxis.

Acute poisoning includes minisoning, moderate poisoning and severe poisoning. The minor poisoning can be defined by th disturbance in the coagulation. The mModerate poisoning can be defined by the symtoms such as bleeding of gums, blood in the faeces, haematuria and haematomata. Severe poisoning can be defined bythe massive haemorrhage which means internal bleeding, cerebrovascular accidents, gastrointestinal blg and retroperitoneal haemorrhage. Flocoumafen can cause acute toxicity in the planktonic algae. The scientific name of this alga algae is Selenastrumcapricornutum. These anticoagulants were toxic to the fishes. Flocoumafen can produce oxicity in the mallard ducks. The toxicity rate can be based on the age of the birds.

The chemicals such as indandiones and hydroxyhydroxyl coumarins can be used as vitamin k antagonists and they are effective in the control of rodents in urban areas. They can also control the rodent populations in the agricultields. Mostly these chemicals can diminish the epoxide cycle of vitamin K_1. The rodent control can be demand several chemicals which that can target even the non- target organisms. They may attack the birds and several. This process can be called as secondary poisoning. The rodenticide regulates certain orders of mammals. They are called as rodentiaRodentia. The warm -blooded vertebrates such as mammals and birds can share similar physiology. Some of the other ertebrates such as fishes, amphibians and reptiles can have different physiologies. Fishes can exhibit a separate physiology patterns. They have a special type of physiology when compared with plants and animals. Mice and rats can be used as model organisms to test the toxic rates of these rodenticides.

The rat poisons can be used in the form of baits. Baits are nothing but the palatable grains. They will be in the form of pastes and this will attract pests. The safety of humans and animals is basedn the toxicity of the rodenticides being used. These chemicals can be ingested by several non-target species. This process is calledas bioaccumulation. There will be an increase in the chemical concentration in the bodies of non- target species. The first generation toxicants are mostly non- coagulants and they are safer than the second -generation anticoagulants. There are reports of the accidental ingestion by the children and adults.

Some of the chemicals which can act as rodenticides are thallium sulfate, zinc phosphide, yellow phosphorus, crimidine, strychnine, fluoroacetamide and sodium fluoroacetate. These chemicals can cause severe pong and it can be acute in nature. These chemicals can be considered as extremely toxic in nature. Out of all the rodenticides known till to date, cholecalciferol can be considered toxic. The red squill, indandiones and anticoagulants can be considered as less harmful than other rodenticides. They create less damage to the non-target species. These chemicals may get accumulated in the liver tissues of humans. Sodium fluoroacetate and strychnine can be used to regulate the coyotes. Indandiones, coumarins and warfarin are used in the United States. They are classified as anticoagulants. Some of these anticoagulants can be classified as first -generation and second- generation anticoagulants.e second -generation coagulants which are toxic than the first generation coagulants can be called as superwarfarins. They can be harmful tothe human beings. Some of the second -generation anticoagulants are difenacoum, brodifacoum and bromodiolone. The chemicals such as indandiones and coumarins can disrupt the hepatic systems. Bcause of this disruption, the blood clotting factors will be damaged. The blood clotting factors will be vitamin K dependent. Some of the bloods clotting factors are II, VII, IX and X. The predominant effect seen in this class is anti-prothrombin effect. This effect can be important to assess and detect cases of clinical poisoning. These agents will enhance the permeability in the capillaries. Due to this, there will be intense hemorrhage. This effect can kill several rodents within days it weeks. The vitamin K dependent clotting factors can be for long periods and so once these rodents ingest the baits there will be susceptible for to the effects of these rodenticides.

To determine the toxic effects of the indandiones and coumarins, the prothrombin timr PT can be used as a marker. Some of the chemicals such asfencaoum and brfacoum can be classified as toxic compounds. They can be toxic tothe non-target organisms also. These chemicals can also affectthe humans even inthe lower doses.odifacoum can be considered as second -generation anticoagulant. It can be classified as a highly toxic chemical which can cause severe damage tothehumans. They can exhibit long half-life periods. There will be long -lasting symptoms due to these rodenticides. These symptoms may occur even after a e exposure. There will be symptomatic bleeding on exposure to indandiones via the dermal route. Some of the indandiones can produce signs related with to the cardiopulmonary and neurological systems. Because of these symptoms, there will be internal hemorrhage too. Some of the symptoms which are crucial to understand are seizures, unconsciousness, confusion and headache. Therbe intracranial hemorrhage which are symptoms associated with the problems in the central nervous systems. The indanedione and coumarin poisoning can be due to enhanced levels of PT. This yields diminished concentrations of plasma prothrombin. Generally, it will take 2 days to show the symptoms.

Yellow phosphorus can be classified as a corrosive agent. This chemical can damage the tissues especially skin tissues. The gastrointestinal tract can also get damaged on exposure to these chemicals. White phosphorous is being used in the fireworks and ammunition. This chemical can be toxic in nature. On exposure, the skin will get bt. In some cases, we can expect third-degree burns. Post ingestion, we can expect symptoms such as mucosal injuries. The mortality rate on exposure tothe phosphorus is about 50%. Zinc phosphide can be used as rodenticides and they are less corrosive in nature. They mainly influence the mucous membranes and skin. The two important phosphides which can be used as rodenticides are aluminum phosphide and zinc phosphide. These chemicals canbe used as fumigants. They are toxic in nature. Phosphide will release phosphine on exposure to the acids present in the gastrointestinal tract. If humans gets exposed to these chemicals, there will be symptoms such as coma, convulsions, delirium, hemorrhage and jaundice.

Thallium sulfate can be absorbed by the skin and gut tissues. It can be absorbed by the liver and kidney tissues. The thallium ions will be presethe bloodstream predominantly. Thallium can be capable of competing with the intracellular K and so the enzyatic ill be altered or disturbed. These This chemical can stay in the bloodstream for atleast 2 days.

Red squill can be considered as a rodenticides and its use is used from the ancient ages. It can be extracted frol cabbages. These cabbages can be grown in the Mediterranean regions. The primary properties and symptoms are due to disturbances in the cardioactive gylcosidesglycosides. The non-target species may not be affected by this chemical. This chemical can be used as a nauseant. Most of the animals can vomit on exposure to these chemicals. Even after frequent vomiting, the toxicity will be retained. The glycosides may not be absorbed in the gut regions. The glycosides may be excreted after several modifications.

Cholecalciferol can be defined as an activated form of the vitamin D molecules. Vitamin D_3 molecule can be considered as crucial in this case. It can be used as effective rodenticides. This can become toxic due to the overutilization of multivitamins. Due to the toxicity of this chemical, kidneys and liver will get affected. Some of the symptoms include nasea, he, weakness and fatigue. This chemical can cause severe damage to the target and non-target organisms.

Bromethalin can be considered as a rodenticides which can beis neurotoxic in nature. It will impair the process of oxidative phosphorylation. Due to this activity, the production of ATP molecules will be lowered andnce the target species will die soon. The potassium/ sodium chanwill be destroyed and hence there will be edema in the cerebral tissues. This will increasethe intracranial pressures. Some of the symptoms which are related with to this chemical poisoning are cerebral edema, coma and mental status variations. If there is an enhanced calciuhe bloodstream, we can define that it is the case of cholecalciferol intoxication.

Rodents are responsible for the damages incurred in the infrastructure industriesindustries. The insulations and chemicals used to control the rodents can be costlier. These rodents can attract predators and the chemicals which are being used to kill the rodents can alsill their predators eventually. The rodents can attract foxes and because of this the poultry and young pigsl be lost. There will be diminished productivity inthe farm animals. There will be noticeable losses in the feed. The rodents can be trapped with the snap-traps. Lot A lot of research has been done to regulate the populations of Norway rats and common rats (few references). This process was done by rodenticides which were classified as anticoagulant rodenticides.

The chemicals such as bromadiolone, difenacoum, difethialone, flocoumafen and brodifacoum can be classified as SGAR or second generation anticoagulant rodenticides. The sugarcane crop can be influenced by the attack of rodents. The damage can be ass by the burrowing behavior, feeding style and nibbling. It is estimated that about 30% of the loss in yield of sugarcane can be affected due to rodent infestation. High A high rate of damage in the sugarcane crop is due to the food source. This problem is dominant in the subtropical and tropical areas.

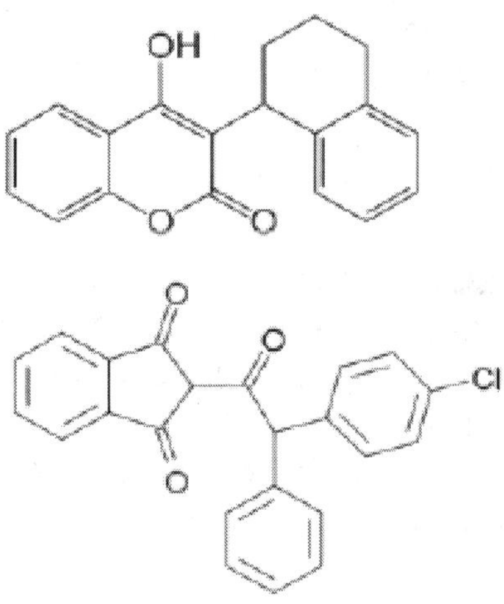

Figure 6.19. Anti-coagulant rodenticides; Coumatetralyl (Top), Chlorophacinone (Bottom).

Source: https://tukes.fi/documents/5470659/6372697/Prevalence+of+anticoagulant+rodenticides+in+non-target+predators+and+scavengers+in+Finland/c901a258-1927-4247-8dea-e37cdfdfb4ab/Prevalence+of+anticoagulant+rodenticides+in+non-target+predators+and+scavengers+in+Finland.pdf

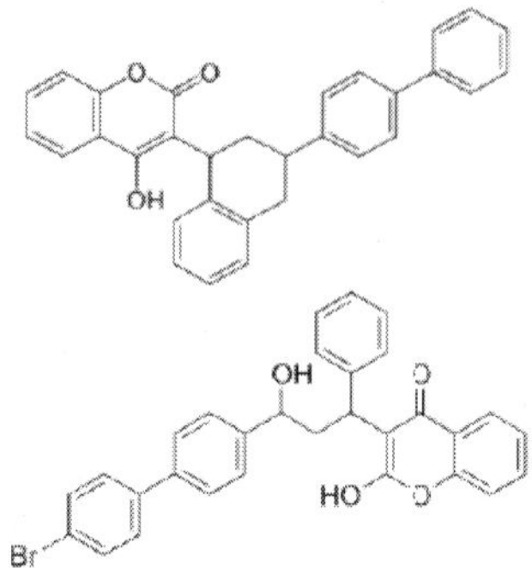

Figure 6.20. Anti-coagulant rodenticides; Difenacoum (Top), Bromadiolone (Bottom).

Source: https://tukes.fi/documents/5470659/6372697/Prevalence+of+anticoagulant+rodenticides+in+non-target+predators+and+scavengers+in+Finland/c901a258-1927-4247-8dea-e37cdfdfb4ab/Prevalence+of+anticoagulant+rodenticides+in+non-target+predators+and+scavengers+in+Finland.pdf

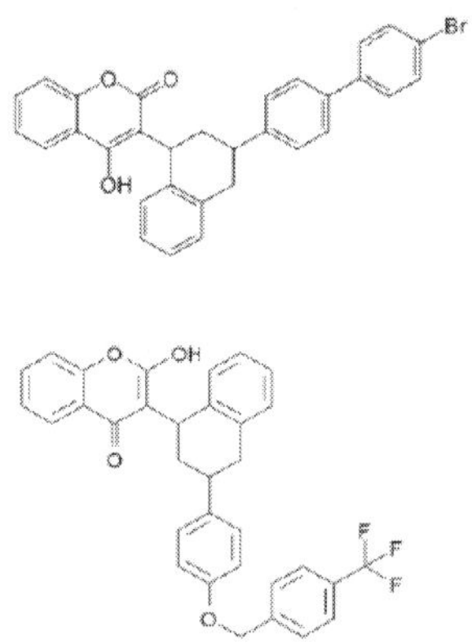

Figure 6.21 Anti-coagulant rodenticides; Brodifacoum (Top), Flocoumafen (Bottom)

Source: https://tukes.fi/documents/5470659/6372697/Prevalence+of+anticoagulant+rodenticides+in+non-target+predators+and+scavengers+in+Finland/c901a258-1927-4247-8dea-e37cdfdfb4ab/Prevalence+of+anticoagulant+rodenticides+in+non-target+predators+and+scavengers+in+Finland.pdf

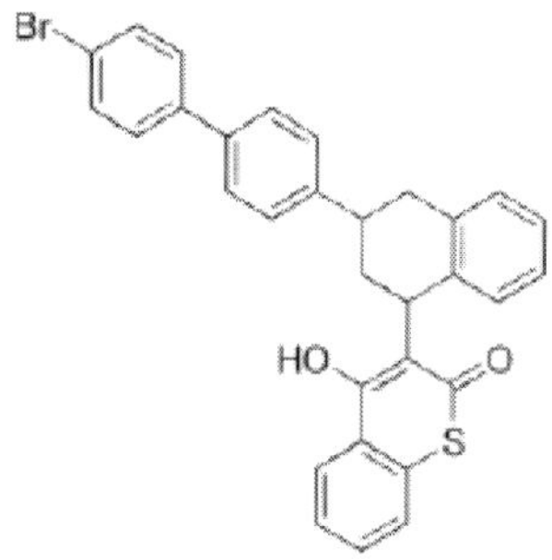

Figure 6.22. Anti-coagulant rodenticides; Difethialone.

Source: https://tukes.fi/documents/5470659/6372697/Prevalence+of+anticoagulant+rodenticides+in+non-target+predators+and+scavengers+in+Finland/c901a258-1927-4247-8dea 37cdfdfb4ab/Prevalence+of+anticoagulant+rodenticides+in+non-target+predators+and+scavengers+in+Finland.pdf

7

Microbial pesticides

CONTENTS

7.1 INTRODUCTION

Microbial pesticides can be used to regulate the pests effectively. They maintain chemical, parasitic and predatory relationships with the hosts. The process in which the microbial pesticides control the progress of the parasites can be caldas biological control. Most of the biopesticides can be considered as a form of pesticiich that can be basenthe natural products and microbes. These biopesticides can include microbes and this type of pesticides can be lledas microbial pesticides. The pesticides which can be produced by the plants itself to regulate the parasites or pests can be called as plant- incorporated protectants. The organisms which can be used to control the pests can include nematodes, fungi and other microbes. The Microbial pesticides can include vhogenic entomopathogenic fungi and bacteria. The entomopathogenic nematodes can be classified as microbial pesticides. Some of the chemicals can be defined as bio-derived chemicals. The bio-derived chemicals can be classified into four groups. They are neem oil, rotenone, pyrethrum and other essl oils . Theand they are called as plant -incorporated protectaprotectants. They contain geneterials which that are present in the genetic makeup itself. These crops can be called as GM crops or genetically modified crops.

Entomopathogenic bacteria	
Lepidoptera	*Bacillus thuringiensis* subspecies kurstaki, *B. thuringiensis* subspecies aizawai
Coleoptera: Scarabaeidae, *Popillia japonica*	*B. thuringiensis* subspecies japonensis, *Paenibacillus popilliae*
Entomopathogenic fungi	
Coleoptera (Scarabaeidae)	*Beauveria brongniartii*
Hemiptera, Thysanoptera	*Conidiobolus thromboides* Acari
Hemiptera	*Aschersonia aleyrodis*, *Lecanicillium longisporum*
Coleoptera, Diptera, Hemiptera, Isoptera	*Metarhizium anisopliae* sensu lato
Lepidoptera	*Nomuraea rileyi*

Figure 7.1. Microbial pesticides.

Source: Source:https://www.researchgate.net/publication/321846961_Microbial_Pesticides_Development_Prospects_and_Popularization_in_India/link/5b98d79d299bf14ad4d14561/download

These microbial pesticid can influencethe photosynthesis mechanisms. The plant growth cbe controlled bythe microbial pesticides. Mostly these pesticides can help the plants against several plant pests. There arral chemicals which that can be produced by the plants and they can protect the plants from several pests. These chemicals can be defined as antifeedants. Mostly these chemicals can be classified as renewable and biodegradable alternatives. RNA interference is a widely used technique to regulate the pests which are attacking several plant species all over the world. Bacillus thuringiensis can be used to regulate the growth of Diptera, Coleoptera and Lepidoptera. The toxins which are produced from thisicrobe can be calledasBttoxin. This can be used in the process of genetic engineering. Some of the microbial control agents are active on entomopathogenic fungi and some other useful nematodes. Some of the plant and fungal extracts are semiochemicals and insect pheromones. Spinosad can be considered as fermentation products.

The microbial pecides can be crucial inthe integrated pest management. The biocontrol agents can work as insecticides, fungicides and herbicides. The fungicides and chemicalcides have been used since for a long time.

S. no.	Name of biopesticide
1	*Bacillus thuringiensis var. kurstaki*
2	*Bacillus thuringiensis var. israelensis*
3	*Bacillus thuringiensis var. galleriae*
4	*Bacillus sphaericus*
5	*Trichoderma viride*
6	*Pseudomonas fluorescens*
7	*Trichoderma harzianum*
8	NPV of *Helicoverpa armigera*
9	NPV of *Spodoptera litura*
10	*Beauveria bassiana*
11	*Neem based pesticides*
12	*Cymbopogon*

Figure 7.2. Biopesticides.

Source: https://www.researchgate.net/publication/321846961_Microbial_Pesticides_Development_Prospects_and_Popularization_in_India/link/5b98d79d299bf14ad4d14561/download

Advantages	Characterization
Cost-effectiveness	Costlier but with less number of applications
Persistence and residue effect	Mostly biodegradable and self-propagating
Knockdown effect	Delayed
Handling and bulkiness	Bulky: carrier based using talc powder
	Easy: liquid formulations
Pest resurgence	Less
Resistance	Less prone
Effect on beneficial flora	Less harmful on beneficial insects (pest specific)
Waiting time	Almost nil
Nature of control	Preventive as well as killing of pests
Shelf life	Less/dependent on storage conditions

Figure 7.3. Benefits of Biopesticides.

Source: Source:https://www.researchgate.net/publication/321846961_Microbial_Pesticides_Development_Prospects_and_Popularization_in_India/link/5b98d79d299bf14ad4d14561/download

The fungicides and the chemical pesticides have been used for tterment of agricu since for a long time. The cChemical pesticides can pollute groundwater and drinking water. They can damage all the water sources nearby and hence the entire ecological properties will be at stake. It is important to solve this issue so that the remaining natural sources may not be affected. Bacillus thuringiensis can be active against coleopteran and lepidopteran members. Biopesticides can be o regulate several pests which that can damage crops.

They can be in the form of microbes, microbial herbicides and microbial pesticides. We call some of se microbes which can regulatethe insect populations as bioinsecticides. The microbial herbicides can kill the weeds and the crops will not be affected. The secretions generated by the microbes can be used to regulate the viruses, fungi and bacteria. These agents can be used to regulate the diseases, weeds, pests and insects. The microbial pesticides can be bacteria, fungi, protozoa and viruses.

7.2 BACTERIAL PESTICIDES

The bacterial pesticides can be used to regulate the pests which can damage the cropThe Pseudomonas spp. can regulatethe plant diseases. This bacterium can produce antibiotics and siderophores. It can colonize the root systems. Because of the aggressive growth of this bacterium, the local microbes in the rhizosphere will be damaged and hence the plants will remain unprotected. Another bacterium which can protect the plant species from the pests is Pseudomonas fluorescence. This bacterium can be considereas a crucial agent to control microbiales attack on the plants. P.cepacia can be used as an antagonist. This bacterium cancontrol several pathogens. It can produce pyrrolinitrin and this chemical can be considered as an antibiotic. When iron content is limited, there will bsecretion of the siderophores. This will help in obtaining the iron molecles from the soils. This bacterium can uses the siderophores in order to procure large amounts of iron molecules. Due to this process, there will be competition among the pests and hence iron will be scarce for several microbes. Some of the other species which can be considered as crucial in combating the pests which cause damage to the crops are P. alcaligen, P.acidovorans and P. maltophila. The seeds must be inoculated with these agents so that they can be able to control the eases. P.fluorescense can be used to combatthe plant diseases. Due to its application and inoculation on the seeds, the plants can become resistant towards the leaf roller gypsy moth locs and berry borers. These species can attackthe vegetable crops and rice crops.

PATHOGEN	PRODUCT NAME	HOST RANGE	USES AND COMMENTS
BACTERIA			
Bacillus thuringiensis var. *kurstaki* (*Bt*)	Bactur®, Bactospeine®, Bioworm®, Caterpillar Killer®, Dipel®, Futura®, Javelin®, SOK-Bt®, Thuricide®, Topside®, Tribactur®, Worthy Attack®	caterpillars (larvae of moths and butterflies)	Effective for foliage-feeding caterpillars (and Indian meal moth in stored grain). Deactivated rapidly in sunlight; apply in the evening or on overcast days and direct some spray to lower surfaces or leaves. Does not cycle extensively in the environment.
Bacillus thuringiensis var. israelensis (*Bt*)	Aquabee®, Bactimos®, Gnatrol®, LarvX®, Mosquito Attack®, Skeetal®, Teknar®, Vectobac®	larvae of *Aedes* and *Psorophora* mosquitoes, black flies, and fungus gnats	Effective against larvae only. Active only if ingested. *Culex* and *Anopheles* mosquitoes are not controlled at normal application rates.. Does not cycle extensively in the environment.
Bacillus thuringiensis var. *tenebrinos*	Foil® M-One® M-Track®, Novardo® Trident®	larvae of Colorado potato beetle, elm leaf beetle adults	Effective against Colorado potato beetle larvae and the elm leaf beetle. Like other *Bts*, it must be ingested. It is subject to breakdown in ultraviolet light and does not cycle extensively in the environment.
Bacillus thuringiensis var. *aizawai*	Certan®	wax moth caterpillars	Used only for control of was moth infestations in honeybee hives.
Bacillus popilliae and *Bacillus lentimorbus*	Doom¨, Japidemic¨,® Milky Spore Disease, Grub Attack®	larvae (grubs) of Japanese beetle	The main Illinois lawn grub (the annual white grub, *Cyclocephala* sp.) Is NOT susceptible to milky spore disease.
Bacillus sphaericus	Vectolex CG®, Vectolex WDG®	larvae of *Culex, Psorophora,* and *Culiseta* mosquitos, larvae of some *Aedes* spp.	Active only if ingested, for use against *Culex, Psorophora,* and *Culiseta* species; also effective against *Aedes vexans.* Remains effective in stagnant or turbid water

Figure 7.4 BacterialBiopesticides.

Source: https://www.intechopen.com/predownload/44118

Bacillus thuringiensis can be classified as a spore- forming bacteria. They are gram positive rods. They can produce crystal proteins which can be secreted as parasporic inclusions.

These parasporic inclusions can bseen in the vegetative cells which can be calledas spores. The crystal proteins can be in the form of protoxins. Due to the activation of the protoxins by the enzymes and pH in the gut of the insects,

They get transformed into active toxin molecules. The toxins can kill the insects once they become active. Some of the target pests for this bacterium are caterpillar pests.

Some of these pests can be classified as dipeteran and lepidopteran mehis bacterium can also attack the cockroaches, bulwormsbollworms, backflies, anopheles and culex mosquitoes. These pests can damage vegetable crops, citrus, maize, tobacco and cotton crops.

FUNGI			
Beauveria bassiana	Botanigard®, Mycotrol®, Naturalis®	aphids, fungus gnats, mealy bugs, mites, thrips, whiteflies	Effective against several pests. High moisture requirements, lack of storage longevity, and competition with other soil microorganisms are problems that remain to be solved.
Lagenidium giganteum	Laginex®	larvae of most pest mosquito species	Effective against larvae of most pest mosquito species; remains infective in the environment through dry periods. A main drawback is its inability to survive high summertime temperatures.

Figure 7.5. Fungal Biopesticides.

Source: https://www.intechopen.com/predownload/44118

PROTOZOA			
Nosema locustae	NOLO Bait®, Grasshopper Attack®	European cornborer caterpillars, grasshoppers and mormon crickets	Useful for rangeland grasshopper control. Active only if ingested. Not recommended for use on a small scale, such as backyard gardens, because the disease is slow acting and grasshoppers are very mobile. Also effective against caterpillars.

Figure 7.6. Protozoan Biopesticides.

Source: https://www.intechopen.com/predownload/44118

VIRUSES			
Gypsy moth nuclear plyhedrosis (NPV)	Gypchek® virus	gypsy moth caterpillars	All of the viral insecticides used for control of forest pests are produced and used exclusively by the U.S. Forest Service.
Tussock moth NPV	TM Biocontrol-1®	tussock moth caterpillars	
Pine sawfly NPV	Neochek-S®	pine sawfly larvae	
Codling moth granulosis virus (GV)	(see comments)	codling moth caterpillars	Commercially produced and marketed briefly, but no longer registered or available. Future re-registration is possible. Subject to rapid breakdown in ultraviolet light.

Figure 7.7. Virus Biopesticides.

Source: https://www.intechopen.com/predownload/44118

ENTOMOGENOUS NEMATODES			
Steinernema feltiae (=*Neoaplectana carpocapsae*) *S. riobravis*, *S. carpocapsae* and other *Steinernema* species	Biosafe®, Ecomask®, Scanmask®, also sold generically (wholesale and retail), Vector®	larvae of a wide variety of soil-dwelling and boring insects	*Steinernema riobravis* is the main nematode species marketed retail in the U.S. Because of moisture requirements, it is effective primarily against insects in moist soils or inside plant tissues. Prolonged storage or extreme temperatures before use may kill or debilitate the nematodes.
Heterorhabditis heliothidis	currently available on a wholesale basis for large scale operations	larvae of a wide variety of soil-dwelling and boring insects	Not commonly available by retail in the U.S.; this species is used more extensively in Europe. Available by wholesale or special order for research or large-scale commercial uses.

Figure 7.8. Entomogeneous nematode Biopesticides.

*Source:*https://www.intechopen.com/pload/44118

There are some other bacterial species which that can control the pests. They are B. lentimorbis, B.popillae, B. sphand B. subtilis. B. lentimorbis can act on several bettlesbeetles. This bacterium can be effective on the culex mosquitos. The bacterium B.subtilis can be used to combat the infection caused by Pythium species in tomatoes. Rhizoctinasolani infection of the tubers can be controlled. Agrobacterium tumefaciens can cause crown gall diseasein several plants especially dicots. This biocontrol agents can also regulate the crown gall producing bacterial species.

7.3 ENTOMOPATHOGENIC FUNGI

B. bronriatiiand B. bassiana can be used to kill the fungal members. Trichoderma an be used to regulate the root pathogens such as Fusaroum Fusarium Some of the other species which can be regulated by this these fungi re Sclerotium rolfsii, R.solani, Rhizoctonia batalicolaandpPythium aphaium. These pathogens produce collar rot, root rot, famping damping off and wilts. These species can infect the vegetable crops and oil seeds. Among these fungal species,T. virideand T.harzianumcan be considered as effective biocontrol agents. The former species can control Fusarium solani. Hirsutellathompsoniican kill rust mites of citrus species. This rust mites can be called as Phylloucoptrulaoleivora. Their main target is the citrus crops which can be grown in several habitats. This biocontrol fungal species can also act against citrus bud mite or Eriophyessheldoneand Eryophyesguerreronisor coconut. Verticillium lecanii can be classified as non-fasticidous fastidious fungi. This species can kill aphids and seal insects. We can use the conidia of this species to control the growth of aphids, seals and even wstospores. Numeraearileyi can be classified as ascomycteous ascomycetous fungi. This species can act as an enthogenic agent. can veral caterpillar pests which that can attack on the valvet velvet beans, soybean, clover and cabbage.

Figure 7.9 Bacillus thuringiensis var. kurstaki (Bt)

Source: https://www.intechopen.com/predownload/44118

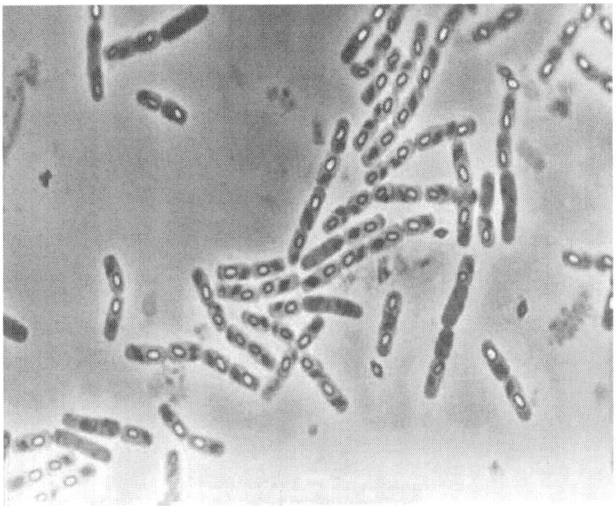

Figure 7.10. Bacillus thuringiensis var. israelensis (Bt).

Source: https://www.intechopen.com/predownload/44118

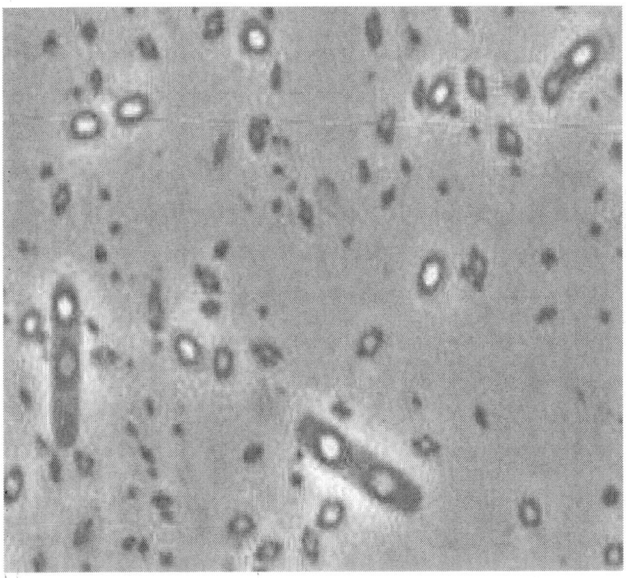

Figure 7.11. Bacillus thuringiensis var. tenebrinos.

Source: https://www.intechopen.com/predownload/44118

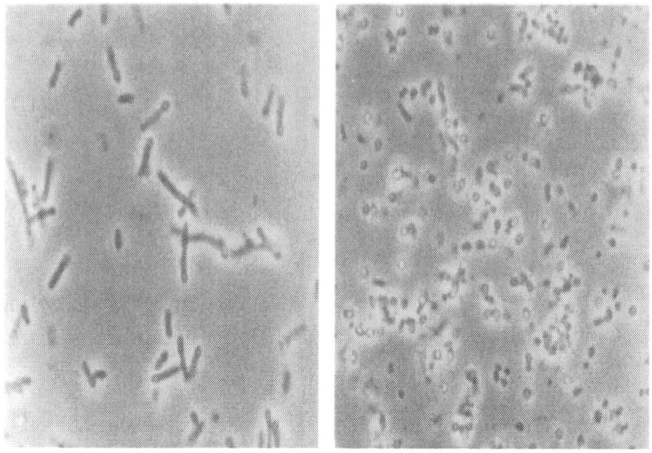

Figure 7.12. Bacillus thuringiensis var. aizawai.

Source: https://www.intechopen.com/predownload/44118

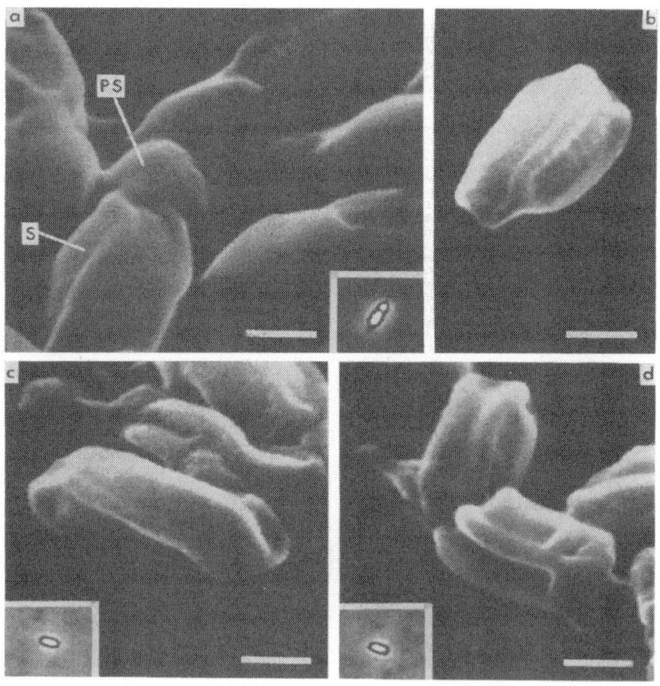

Figure 7.13. Bacillus popilliae(a-c) andBacillus lentimorbus (d)-Spores.

Source: https://www.intechopen.com/predownload/44118

Figure 7.14. Bacillus popilliae and Bacillus lentimorbus killing beetles.

Source: https://www.intechopen.com/predownload/44118

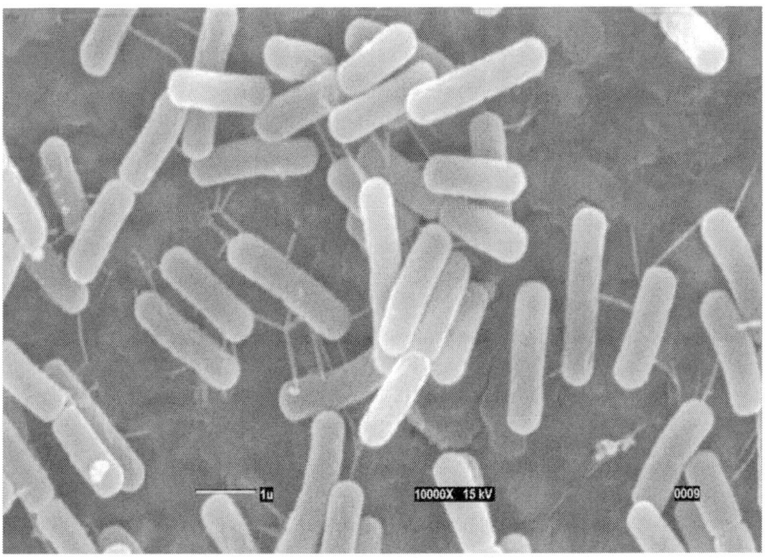

Figure 7.15. Bacillus sphaericus.

Source: https://www.intechopen.com/predownload/44118

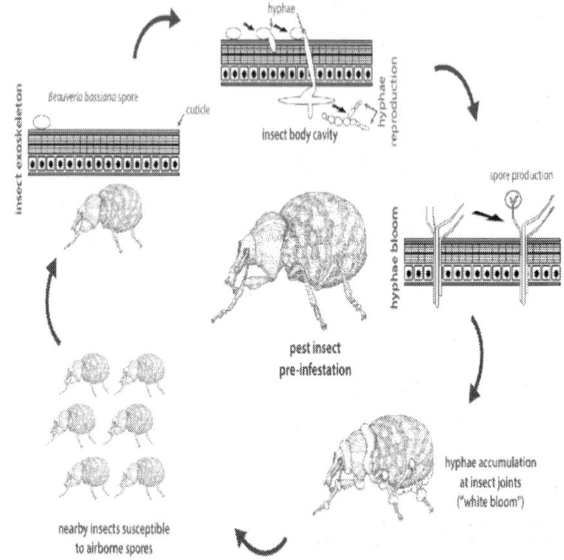

Figure 7.16. Beauveria bassiana.

Source: https://www.intechopen.com/predownload/44118

Figure 7.17. Beauveria bassiana (Killing pests).

Source: https://www.intechopen.com/predownload/44118

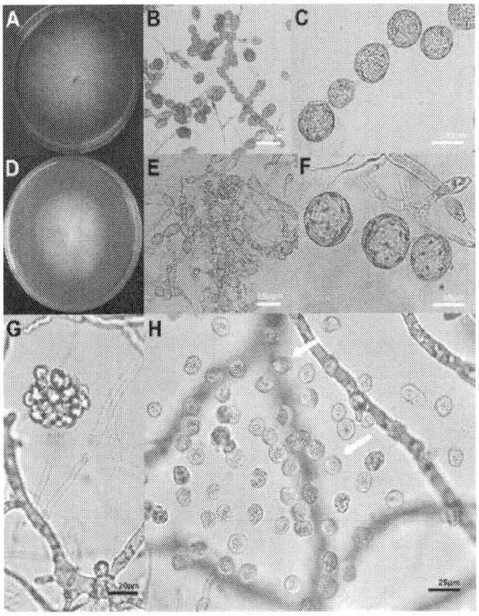

Figure 7.18. Lagenidiumgiganteum.

Source: https://www.intechopen.com/predownload/44118

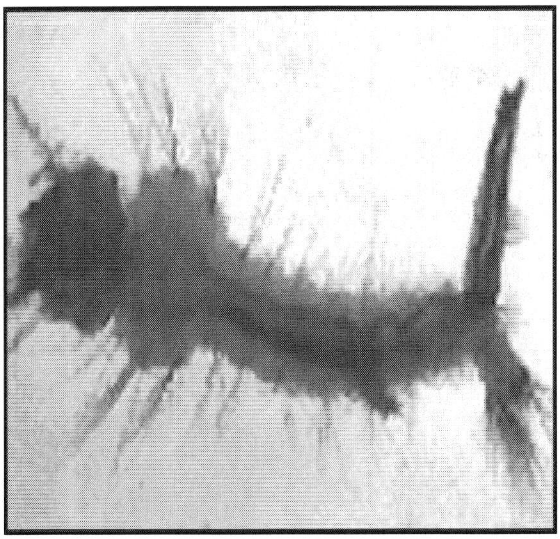

Figure 7.19. Lagenidiumgiganteum (Infected mosquito larvae).

*Source:*https://www.intechopen.com/predownload/44118

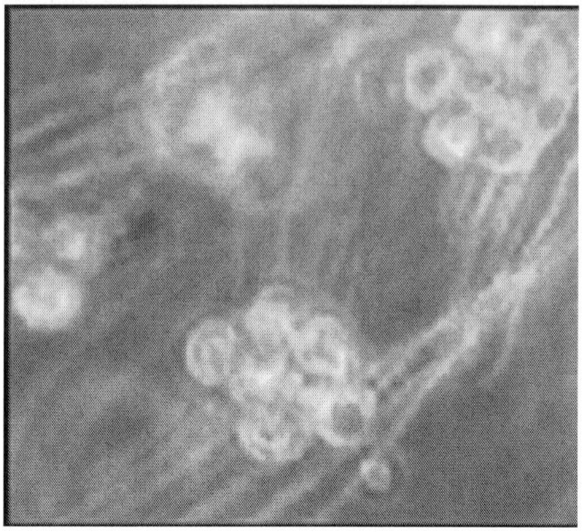

Figure 7.20. Lagenidiumgiganteum (oospores).

Source: https://www.intechopen.com/predownload/44118

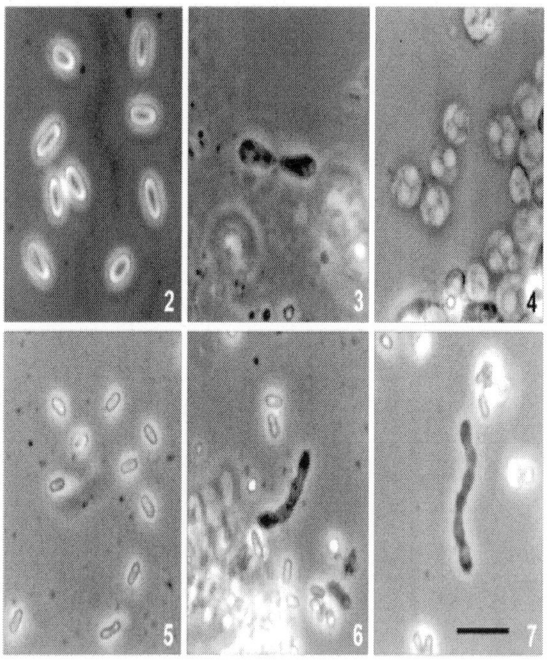

Figure 7.21. Nosema locustae (developmental stages).

Source: https://www.intechopen.com/predownload/44118

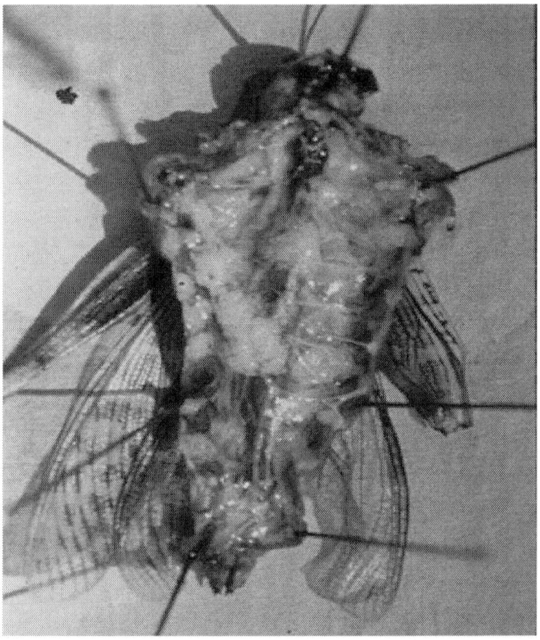

Figure 7.22. Nosema locustae (infecting pest).

Source: https://www.intechopen.com/predownload/44118

Figure 7.23. Gypsy moth nuclearpolyhedrosis (NPV).

Source: https://www.intechopen.com/predownload/44118

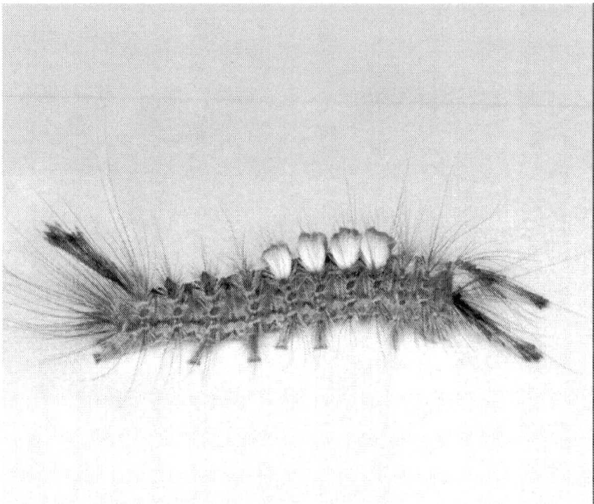

Figure 7.24. Tussock moth NPV (caterpillar).

Source: https://www.for.gov.bc.ca/rsi/foresthealth/pdf/dftm%20npv%20 virus%20information%20sheet.pdf

Figure 7.25. Tussock moth NPV.

Source: https://www.for.gov.bc.ca/rsi/foresthealth/pdf/dftm%20npv%20 virus%20information%20sheet.pdf

Figure 7.26. Pine sawfly NPV.

Source: https://www.for.gov.bc.ca/rsi/foresthealth/pdf/dftm%20npv%20 virus%20information%20sheet.pdf

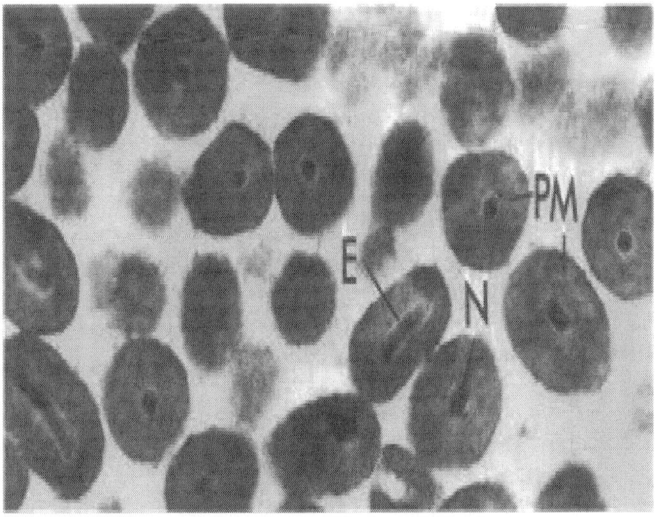

Figure 7.27. Codling moth granulosis virus (GV).

Source: https://www.for.gov.bc.ca/rsi/foresthealth/pdf/dftm%20npv%20 virus%20information%20sheet.pdf

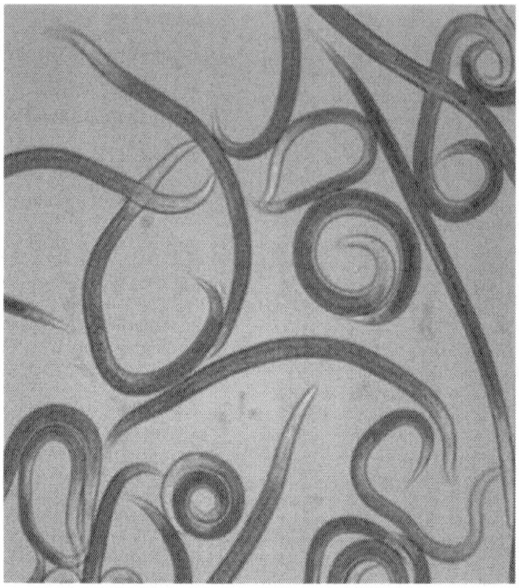

Figure 7.28. Steinernemafeltiae.

Source: https://www.for.gov.bc.ca/rsi/foresthealth/pdf/dftm%20npv%20 virus%20information%20sheet.pdf

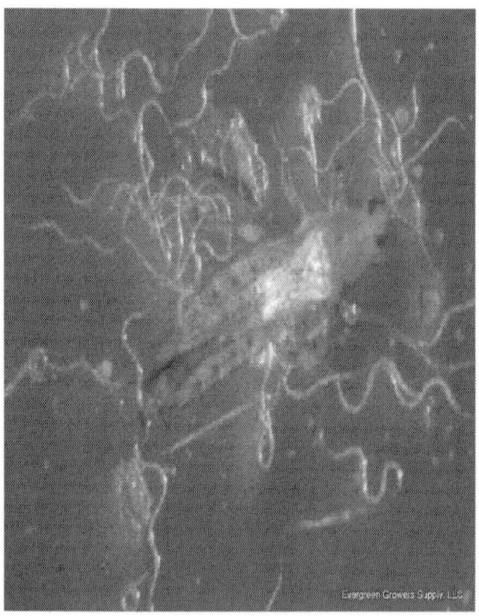

Figure 7.29. Steinernemafeltiae infecting flies.

Source:https://www.for.gov.bc.ca/rsi/foresthealth/pdf/dftm%20npv%20 virus%20information%20sheet.pdf

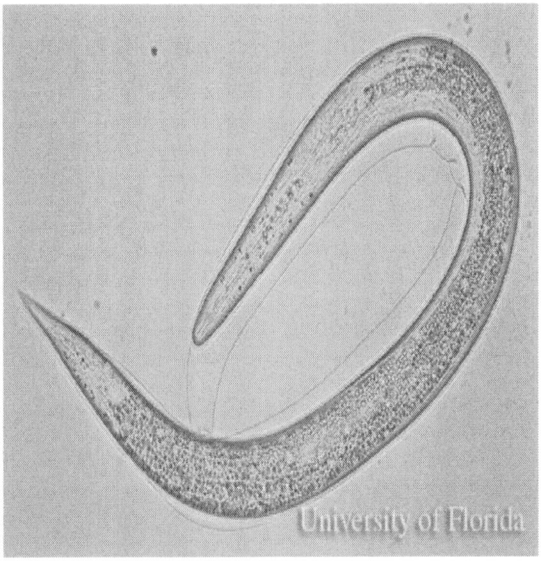

Figure 7.30. Riobravis, S. carpocapsae.

Source:https://www.for.gov.bc.ca/rsi/foresthealth/pdf/dftm%20npv%20 irus%20information%20sheet.pdf

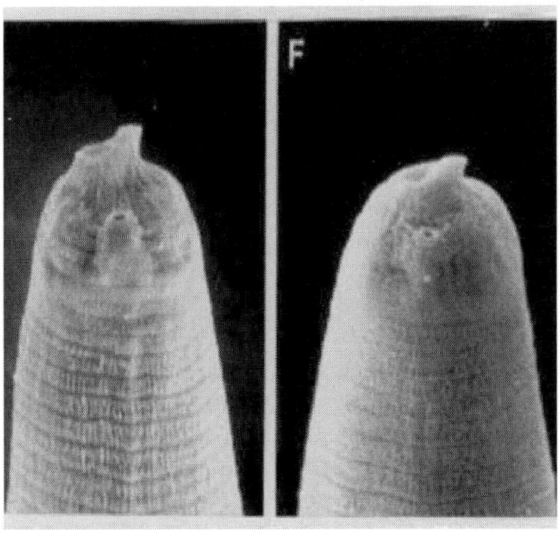

Figure 7.31. HeterorhabditisHheliothidis.

Source:https://www.for.gov.bc.ca/rsi/foresthealth/pdf/dftm%20npv%20 virus%20information%20sheet.pdf

8

Biochemistry of Pesticides

CONTENTS

8.1 INTRODUCTION

The chemicals were usedto regulate the insects from ancient times. This chemical dates back to ancient Rome and Greece. In the year 1896, the rape farmers in France used the Bordeaux mixture. The farmers sprayed this mixture on the weeds and cereals.

In recent decades, there was a gradual improvement in the production of these pesticides.

These pesticides never harmed the crops. There were different inorganic substancesh that were produced in the First World War. Most of the pesticides were produceg in 1913 to be used in the wars.

The organomercury dressing of the seed was first done by Germany during the First World War. The complexity in the chemicals used as pesticides increased due to the identification of different pests and damages they inflict on the crops.

In some instances, Tar oil was used to kthe pests such as aphids. In 1932, the chemical named dinitro orthocresol was patented in France. In same way, Thiram chemical was patented in the United States. Thiram can be classified as dithiocarbamate fungicide.

The second- generation insecticides such as organophosphorus based chemicals evolved during 1950 and 1960. These chemicals proved to be less toxicthe higher animals and humans.

In Japan and Britain, there were experiments done to produce synthetic pyrethroids. These chemicals were resistant to changes caused due to sunlight and were sold to be used as soil pesticides.

Another popular insecticide is the diflubenzuron, which was used to inhibit enzymes needed for the cuticle production. In the year 1960, there was a development of systemic fungicides.

Some of the systemic fungicides are metalaxyl, hexaconazole and benomyl. Metalaxyl can be classified as acylalanine derivative. Hexaconazole can be classified as triazole and Triarimol, ethirimol comes under pyrimidines.

Benomyl can be considered as benzimidazoles. In Britain, during the 1960s, there was excess production of quaternary ammonium compounds. In Switzerland, during the 1960s, there was a production of triazines.

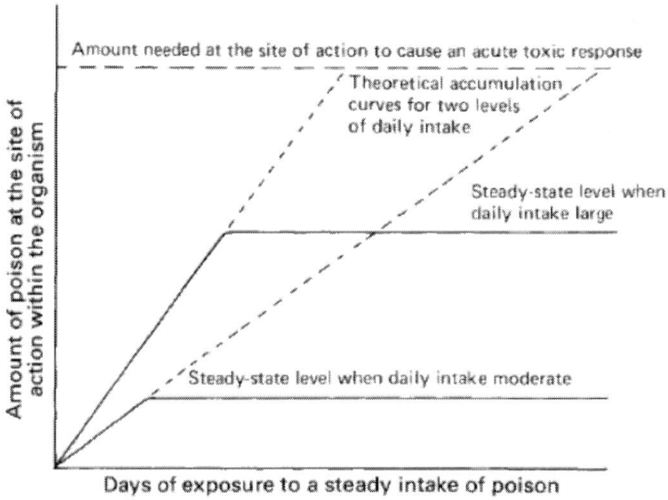

Figure 8.1. Chronic toxicity.

Source: Hassall, K. A. (1990). The biochemistry and uses of pesticides: structure, metabolism, mode of action and uses in crop protection. London: Macmillan.

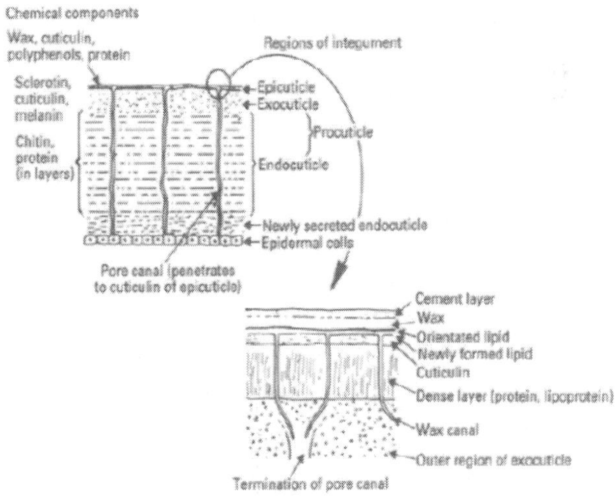

Figure 8.2. Insect integument.

Source: Hassall, K. A. (1990). The biochemistry and uses of pesticides: structure, metabolism, mode of action and uses in crop protection. London: Macmillan.

The intensification of agriculture has demanded excess use of pesticides. The crop rotation process was rarely used in the olden days. The crop rotation system can be useful for pests also. The intensified agriculture demands more area and hence the quota of the pesticides being produced may not be sufficient. In some cases, the pests may generate resistance towards the pesticides. The agriculture system which is in the form of monoculture can be targeted by many pests and diseases. We can consider the monoculture type of agriculture as a metastable system. The presence of weeds, fungi and insects can produce an imbalance in the equilibrium of the agriculture at staitions. IndustrialThe industrial revolution was at its peak after the Second World War. Post the year 1945, several p producing industriefactories came into being. The total expenditure to the farmers on these pesticides rose to $20000 million annually. The use of irrigation, fertilizers and mechanization has enhanced the productivity of the agricultural workers. It is affirmative that chemicals can produce toxic effects. The acute effect can be observed readily. This effect can be visible just with a single dose of operation. Arsenious oxide can p toxic effects which that are more dangerous than sodium chloride.

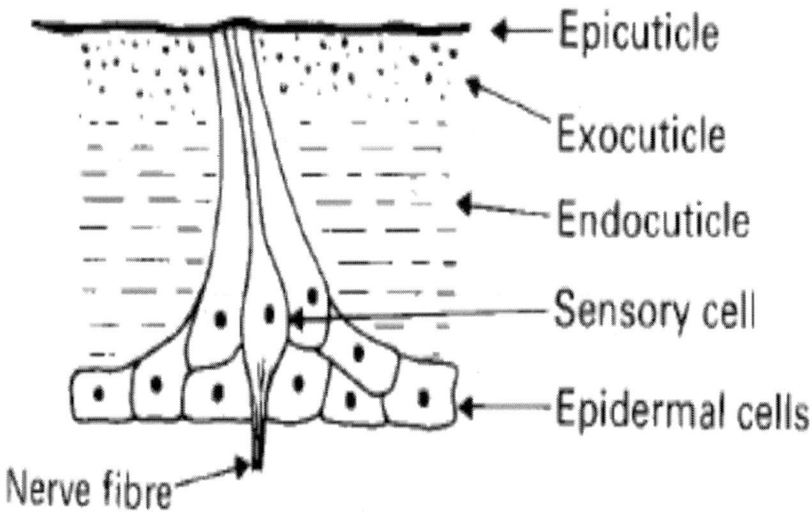

Figure 8.3. Insect cuticle modifications.

Source: Hassall, K. A. (1990). The biochemistry and uses of pesticides: structure, metabolism, mode of action and uses in crop protection. London: Macmillan.

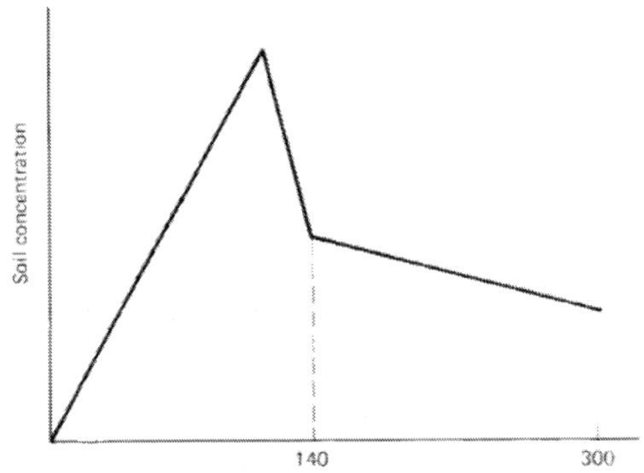

Figure 8.4. Release of toxicants.

Source: Hassall, K. A. (1990). The biochemistry and uses of pesticides: structure, metabolism, mode of action and uses in crop protection. London: Macmillan

In the case of acute toxicity, the physiological and biochemical systems wiamaged at an instant. Chronic The chrc effect can be observed asthe time progresses.

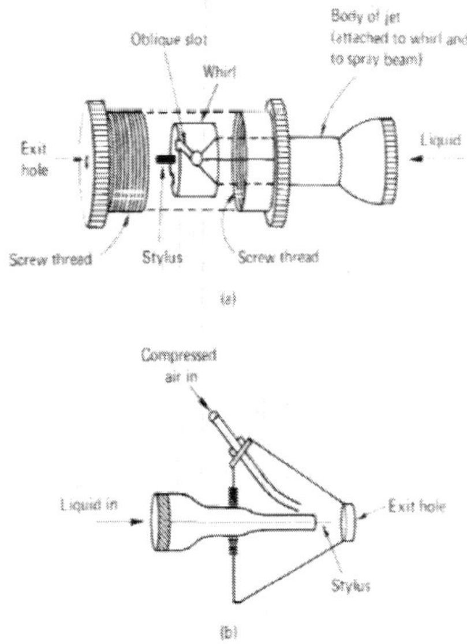

Figure 8.5. Spray delivery systems.

Source: Hassall, K. A. (1990). The biochemistry and uses of pesticides: structure, metabolism, mode of action and uses in crop protection. London: Macmillan

In this case, the chemicals can be used in the form of small doses for long periods. Some of the chronic effects caused due to the use of pesticides are necrosis of kidneys and livers, brain damage and lung cancer. The highest concentration of any pesticides which can cause no damage to the health can be defined as 'threshold daily dose'. This dose can be measured in mg/kg per day. In the case of humans, we can express this dosage in the form of 'acceptable daily intake'. Safety factors must be considered while explaining the ADI or acceptable daily intake and threshold daily intake. These factors can constitute 2 important components. The first component is the interspecies variation. The first component will assume that humans may become more susceptible to the pesticide than the test species. The second compond as intra-speat will be of a statistical naturestatistical. It is affirmative that one species can present more toxicity than others. This toxicity response will be observed among single species.

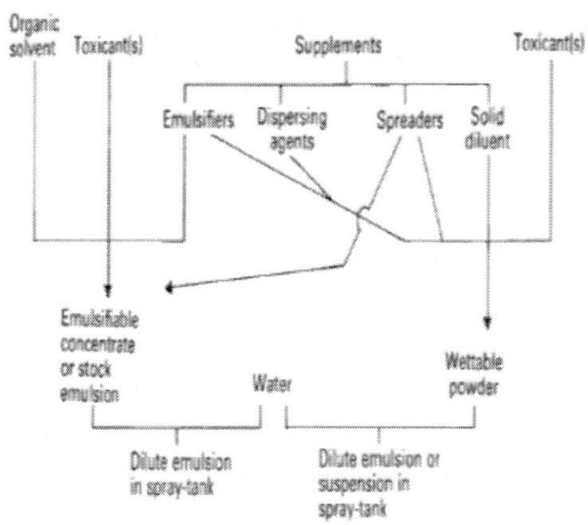

Figure 8.6. Spray delivery systems-components.

Source: Hassall, K. A. (1990). The biochemistry and uses of pesticides: structure, metabolism, mode of action and uses in crop protection. London: Macmillan

The pesticides can produce some mutations which can be harmful in nature. The mutagenic effects mainly disturb the chosomes. This can lead to cancers inthe humans as well as animals. They can function as mutagens and can produce carcinogenicity. In order to evaluate the mutagenic nature of the pesticides, we need to consider the Ames test. It is important to know that the hazard levels of these pesticides. The first level of hazard will rest with the manufacture, storage and transport. In thisaccident occurs inside theindustpowderry, then it will be treated as first -level hazard. This level of hazard can be treated as serious. The second level of hazard can be at the spray tank level. The pesticides must be used in dilute concentrations on crops and this will cause problems to the workers at that place. In many cases, the workers willposed to a specific concentration which that cfunctional in huma The workers which who are exposed tothe pesticides can die and many deaths were reported for the same cause due to accidental ingestion of these chemicals (referennn). The grains and pulses sprayeiththe pesticides can be dangerous tothe human life only if ingested without cleaning properly.

At the time of consumption, if the pesticide concetration is higher, then we can call it as third level hazard. The functional concentrations are just a portion of the pesticide residues. The residual levels can be harmful and the ingestion of these chemicals can lead to chronic toxicity.

The pesticides be sprayed on the leaves in the form of dustpowders and sprays. The cuticle present in the leaves will take up the leaves. The cuticle can be considered as a barrier for fungi and other organisms and when the pesticides are sprayed on it, the pests will eventually die. The cuticle of the leaf can be considered as a direct target. The cuticle barrier must be breached so that the insecticides can reach the site of action. The spore walls and the walls of fungal hyphae can be considered as sites of action for any pesticide. The intermediate target in this process is the soils. The soils which can be enriched with the silicates can be assumed as having negative charges. The soils mostly contain spaces which are filled with air. The cuticle can be waxy in nature and there is much variation observed between soil and cuticle of the leavThe pesticides must be formulated prior toits their application.

The protective layer can be placed exterior to the cellulose walls and in most cases, the cuticle will be merged with the cells in the epidermis. The parenchymatous cells present in the leaves are younger and they are demated by pectic acid. The waxy cuticular can be vary in their composition. The plants are having alkanes and s. There are some specialized structures which that can get pierced form the cuticle layer. They are hydathode channels, trichomes and stomata. The hydathode structures can be forming channels. The water molecules can get out from of these perforations made by the hydathodes. Thegree of penetration can vary with the type ofthe plant species. The absorption rate will depend on the fungicides and insecticides. We can observe persistent contact action and loco systemic action in case of these pesticides. Some of the pesticides can be deposied on the leaf surfaces. Such a type of pesticides can exhibit local contact action. Some of the pesticides can pass through the cuticle and reach the sites of action. Some pesticides can exhibit systemic action in many . The cuticle can be considered as a barrier which that can allow the movement of the pesticides freely. The cuticle is made up of several lipoproteins and theycle the living cells. There are many factors which that can influence the absorption rates of the leaves. Hartley and Graham Bryce in 1980 exined the chemistry underlying the penetration ofthe pesticides via cuticle.

e exterior layer of the insect body can be calledas epidermis.The insects can produce a protective layer called as cuticle. The cuticle layer can protect the insects from several factors. The preparation of the materials which are active on the pests can be cal as formulation. These ingredients can be active on in the living systems. The process of the formulation can be affected by the volatility, flammability and toxicity. The packaging and formulation process needs to be planned perfectly. The active ingredient can be designated as AI can be used in the form of pellets, granules and dustpowder. The active ingredits can be applied in the liquid form which is calledas spray. Water is used as a solvent in many pesticiplications. There are two important formulations which that can be widely used to apply these pesticides. They are named as wettable powders or WP and emulsifiable concentrates or EC. Some of the pesticides can be dissolved in water and they can be used as aqueous solutions. Some of the pesticides can be used as suspension concentrates or SC. They can flow when added with water. The pesticides can flow in the spray tank. The active ingredients can be used in combination with the non-aqueouiquids.

If any active ingredients of the pesticides is a converted to vapor state and then applied, we call itas a fumigant and the process will be termed as fumigation. When fumigated, the vapor produced can enter into the living system without any problem. It iportant to know that the fumigants can work well if the space is encloserapid diffusion is ensured. This is the condition which that limits the use of fumigants in cropction. The fumigants can be used to protect crops which that are being grown in the greenhouses. Some of the pesticides can exist as false fumigants. The false fumigants can be prepared with the help of smoke generators. The oxidant generally used in this process is chlorate or carbohydrate. Some of the active ingredients which can be used via the smoke generators are pyrethroids, azobenzene, captan and lindane.

There are some aerosol preparations which can be used in the form of suspension. Suspensions can be in the form of droplets which appears as mist. We can use fumigation tablets to fumigate the grains. The diffusion of the toxicants which are volatile can be restricted by the soil structures. It is important to use certain specialized devices to fumigate the soils. The fumigation process must be done in intervals. The pesticides which can be used by this method are dibromochloropropane ad dichloropropene. There are some pesticides which exhibitsa low level of toxicity. The lindane and aldrin molecules can regulate the leatherjackets and wireworms. This can be possi with the fumigants. The aerosols can be used to controlthe crop pests. The aerosols can be used to regulate the mosquitoes and flies. The pyrethroid molecules can be used to regulate these pests. We can use the pirimicarb and malathion aerosols to combat the pests. The active ingredient can be dissolved in the solvent and the solution can be applied in the form of a jet. The jet gas can be used to spray on the crops. The mixture of the pesticides will be stored in the vapor state. The mixture can be stored under high pressures. The fluorocarbon can ed in combating the pests and it is regarded as safer. The cCarbon dioxide and butane can be used. These chemicals have no risk of flammty.

Polon has explained the production powders and dustsdusthe pesticides can be linked with thesorptive materials. The transportation costs can be minimized. In order to do that it is important t the formulation process in two stages. The concentrated dustpowder must be shipfirst to the storage house. It is important to dilute the dustpowder at the storehouse. The sorptive diluent will be formed. It will be just about 10% of the actual concentration. The pesticides will be impregnated inside the minerals. ays and silicas can be considered as important minerals which that can besed to impregnate these pesticides. The silicas can compriseof pumice, diatomite and talc. The clays comprise of attapulgite and kaite. The attapulgite can be called as fuller's earth. Tustpowder particles must be less than 50 micrometers. The dustpowder particles which ce used often can measure between 3 to 30 micrometers. The dustpowders which are being deposited can be transported by the rain and wind. Most of the particlhich can be transported are less than 10 micrometers. The dustpowderticles will be finer and it is crl to use an advanced dustpowdering machine to get uniform dustpowder particles so that there may not be clumping.

Water availability is less in most of the countries. In that case can prefer foliage application. The appliances used for dustpowdering can include turbine blowers. They are cheaper and we can easily modify as per our needs. The formulation process can affect the dustpowder dispersion rates. When the pesticides are added with the diluents, the effect will be enhanced. Some of the pesticides which can be used in the form of sprays can be chlorpyrifos and fenitrotin. It is importantknow that the sprays must be operated a higher speeds. The dustpowders of these pesticides must be applied to the soils. The size of the particles can be important so that we can estimate the volatilization rates. There will be an increase in the vapor pressure as the dispersion rates increases. The air spaces present in the soils can be filled with the active ingredients.

The rate of dision can depend on the size of the particles. The pesticide dustpowder can be allowed to sprinkle on the trenches prior to the sowing of seeds andsome cases we can even sprinkle in the holes. The pesticides dustpowder can be scattered on the crops and this will work best to control the root fly of the carrot. The granular formulations can be used to treat several pests at the ground level. Due to this type of treatment, the wastage can be reduced and so the phytotoxicity can be regulated. The granules can be applied to the soil prior to the start of a crop. The application of these granules should be followed by watering and rotavation. The tillage process must be reduced and so the pests can interth these granules. Excess growth of weeds is another problem which that must be considered.riers of these pesticides should be non-reactive and inertin nature. The carrier must be chosen based on the sorptive capacity. We need to know the density and the hardness of the carriers. The carrier should have excellent porous reticulum. The carrier must allow the pesticide to be impregnawith an active ingredient. Some of the supports which are inert are like limestone, vermiculite, kaolin and attapte. Mostly the granules can be produced by the aggregation of the dustpowders with the help of an adhesive. The escape of the toxicity can be decreased with the plastic coating. We can use a thermoplastic matrix. This matrix will have a structure which is having several micropores. The granular formulations can be prepared by adding the active ingredients with the porosity modifiers. The granules can be prepared by adding the AIs with the starch xanthate. In some cases, we can use starch band starch calcium. The granule technology can provide pesticides which that can persist for long periods. The organochlorine compounds can be used and the molecules of this chemical can exist for long periods. The surface of the granules should be coated and so the active ingredient or the AI which is present in the granules can be released slowly into the soil. We can define such a type of toxicant release as a controlled release. We may also define such a type of toxicant release as a delayed release. Some of the soft granules can be prepared from the minerals. We need to fire the hard granules to get this effect. This will help in the dispersion of the toxicants in water.

The granular formulations can range from 0.3 to 1llimeter in diameter. The granules should be added to the soil and they be are not be strictly added in random. The delayed release of the pesticide will be based on the leaching rate of the granules in the presence of water. The rate of the delayed release will be based the depletion of the pesticide. The rate of leaching may be influenced bythe temperature variations. The pesticides must be applied as suspensions or aqueous solutions. Mostly the active ingredients must be added to the organic solven The active ingredients will be dissolved in the oil -based organic solvent. A WP or wettable powder can constitute inert solid and active ingredients. These substances can use other auxiliary compounds such as adjuvants. The wettable powders can be adhering to the leaves and mostly they will be in the oil formulations. Some of the oil formulations include EC or emulsifiable concentrates. They may not be leached or wiped out due to rain. These powders can be solids and they can be used with water. The mixing of water with these solids can occur in the spray tanks. These chemicals can be applied in their dry state.

The EC or the emulsifiable concentrates can be formulations which are oil- based. We can achieve single- phase mixture in this case. This single -phase mixture can produce emulsions. The size of the particles will be lesser than 8 micrometers. These emulsifiable concentrates can have two emulsifiers. The concentration of the mixture can increase if the volume of the pren reduces in the spray tank. The ultra- low volume systems can use five litres liters of fluid for each hectare. It is estimated that 40% of the active ingredients are available to the crop. Water may not be added to these formulations. The spray appears as a mist which is nothing but droplets of the liquid. This type of formulation is not flexible and there will be problems due to evaporation. The droplets may not be distributed equally on the crops. We can alter the dynamics of the spray by adding water or any other liquid which does not decrease the strength of the pesticides. In the spray tanks, there can be two types of supplements which are called as emulsifiers and stabilizers or dispersing agents. These two supplements help to withstand the heterogeneous mixture whilst spraying. We can observe the wetting agent or spreading agent and this can enhance the contact between the surface and the droplets. These supplements will exhibit surfactant properties. The droplet can retain on the surface of the leaves.

The retention time of these droplets can be based on sveral factors. Some of the important factors are the size of the drop, carriers volatility and surface tension. There is a lot of variation in the arrangement of the leaves. The retention capacity can be based on the structure of the leaves. In order to achieve sufficient retention time, it is important that the spreading agent is added iss. Post osition, the emulsions can cleave. There is anoher chemical which that can be help in sticking. This chemical can be called as sticking agent. Some of the sticking agents arynthetic polymers, resins, gums, gelatin, flour and casein. There is a need of for increasing the time before the droplets evaporate. These agents can be called as humectants. The activators and synergists can determine the toxicant's capacity to remain poisonous. The synergists can be used to reduce the activity of the enzyme microsomal oxidase. The activators can be applied to enhance the availability of the active ingredients to the target.

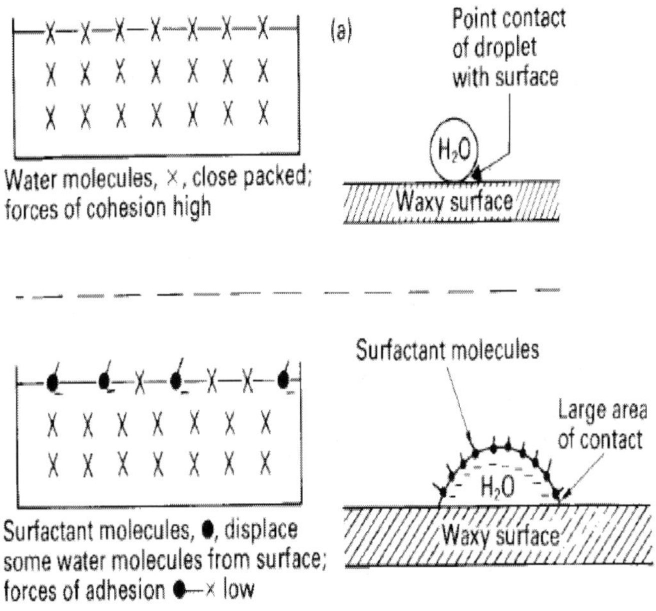

Figure 8.7. Surfactants and surface tension.

Source: Hassall, K. A. (1990). The biochemistry and uses of pesticides: structure, metabolism, mode of action and uses in crop protection. London: Macmillan

The surfactants can be divided into detergents and soaps. The surfactants can be classified into three groups. The anionic surfactants include soaps. These compounds can comprise of non -polar regions. There will be an ionized group at the end of these molecules. The amphipathic ion can have a negative charge. These are many kinds of anionic surface of the important anionic surfactants areydrocarbons and sulphated sulfated alcohols. These chemicals can be articifically artificially produced from the oils and petroleum products. They can react readily with metal ions. The group which comprises of positively charged particles are called as cationic surfactants.

The pyridinium derivatives can be best- suited example for this type. DImethylsilicones can be classified as spreading agents. The presence of an emulsifier or stabilizer can be based on the electrical properties in the colloid.

The spreaders can be functional on impact and they are classified as surfactants. The surfactants comprise of lipophilic and hydrophilic groups. Due to the presence of these groups, there will be balance betweenrges.

8.2 PESTICIDE METABOLISM

Most of the living organisms can exhibit defence defenreactions when there is a threat to their existence. The plants can also resistthe pesticides. It is obvious that when any toxic substance enters an organism and there is no scope of elimination, then it can get accumulated. Due to the aculation of these substances, the living organisms will die.

The accumulation ofthe toxic substances cant in certain adverse reactions at specific sites. There are several factors which that can explain the penetration of the toxic chemicals into the living system. Some of the important factors are biochemical factors, phgical factors and anatomical factors.

There are certain speciald enzymes which that can act on the foreign compounds. These enzymes can act onthe toxic substance and make them less harmful.

These enzymes can also produce certain changes in the toxicants so that they can be soluble in water and hence eliminated by the plant body by the transpiration process or any other processes.

The pesticides cannot be readily dissolved in water. But the hydrolysis and oxidation or in some cases both of the reactions can expose the polar groups on the pesticide molecules.

In some cases, there will be an addition of these groups on the pesticide molecules. These reactions can be called as primary metabolic reactions. There will undergo secondary modifications.

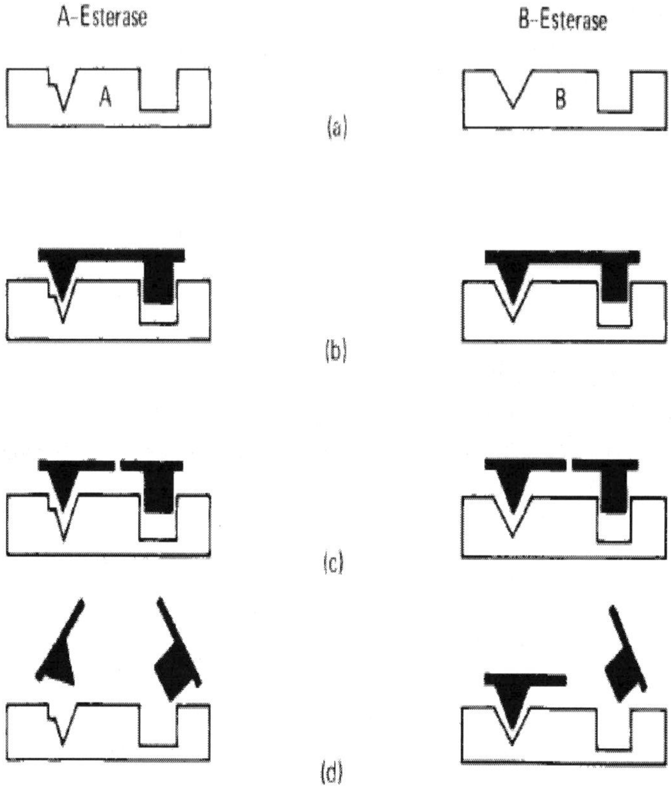

Figure 8.8. A-esterases and B-esterases.

Source: Hassall, K. A. (1990). The biochemistry and uses of pesticides: structure, metabolism, mode of action and uses in crop protection. London: Macmillan

The metabolism process can be crucial to determine the selectivity of the pesticides. In several cases, the selectivity property has been cruc in defining the toxicity farm animals and humans. The metabolism levels can decidethe pesticide persistence inthe living organisms and the environment. The persistence of the pesticides in the atmosphere must be kept at a minimum so that there will be neglignvironmental pollution. The pesticides can undergo certain primary molecular changes which that can make them reactive or neutral and hence metabolized by the microbes. Most of the bal species can use the C skeleton of these pesticides.

There are several mechanisms which that can be useful in determinsistance againstthe pesticides and other chemicals. There are hydrolytic enzymes which are presentimals and plants. They are present in the individual cells. There are hwhich are present in the plasma. In the microsomes, there are certain hydrolases which are attached to the membrane. Most of the enzymes required for the oxidative metabolism can exist in the endoplasmic reticulum of the liver cells. There are many types of microsomal enzymes which can be important in the pesticide metabolism. These enzymes are called as microsomal polysubstrate mono-oxygenases. The pesticides can be reduced dto the enzymes present in the microsomes. Most of the reduction reactions can be present inthe vertebrates. The nitro group present in these pesticides can be reduced to the amino groups. The conjugation processes can occur between the enzymes which are membrane- bound and some cofactors which are dissolved in the cytosol. The glic acid can be linked with the metabolites. Most of the higher animals can have enzymes which that can undergo the conjugation process.

Most of the pesticides are comprised of phosphate linkages, amides and ester linkages. The hydrolases can attack these pesticides. Some of the pesticides which are susceptible to these hydrolases are carbamate insecticides, pyrethroids and organophosphorus based insecticides. The herbicides such as carbamate and urea can be susceptible to the hydrolases. The fungicides such as dinitrophenol and dithiocarbamate can be damaged by the hydrolases. These enzymes may not require coenzymes. Most of the living organisms contain thee hydrolases.

The esterases can be classified into A-esterases and B- esterases. Theerases can disrupt the organophosphates. Paraoxon is an organophosphate based pesticide which that is susceptible to the hydrolases. The B-esterases can be damaged by the organophosphates. Most of the microsomal proteins can be classified as membrane linked esterases. These esterases can be of B-type. These esterases can be present in the pancreatic juice and serum. The carbse enzyme can act on the pesticide malathion and make it detoxified. This enzyme can hydrolyse hydrolyze the 2,4 D esters. These esters can be absorbed by the weeds and dichlorophenoxyacetic acid will be activated. These esterases have developed an affinity towards the carbamates. The pesticides based on carbamates canformed into carbamoyl derivatives. These substances can undergo hydrolysis slowly. The sulphide sulfide molecules can be transformed intes. In some cases, there will be the formation of sulphones. These molecules can get hydrolysed hydrolyzed to produce oximes and carbamic acid derivatives.

The pesticides can have several alkyl groups. Most of the pesticides have O-C_2H_5 and O-CH_3 .These groups can be eliminated due to the degradation caused by the metabolism reactions. Some of the alkoxy groups can be eliminated by the enzyme glutathione transferase. Tde linkage can be similar to the esters. Propanil can be classified as a crucial herbicide which that is selective in nature. This herbicide can be used to regulate the weeds which attack the rice crice plants can have carboxylamidases. This enzyme can protect the rice plant frombeing hydrolysedhydrolyzed. Carboxylamidases can be used to select the amides. These amides can be used as substrates. The carboxylesterase can be used in conferring insecticidal resistance. The epoxide hydrolases can be classified as hydrases and hydratases. These enzymes can be localized in the microsomes. They can be concentrated in the cytosol of the cell. In the liver cells, these enzymes are abundant. These enzymes can damage epoxides. We can observe the peroxidation and auto- oxidation of the pesticides. The epoxides can be active and can undergo oxidation. Due to the oxidation process, the DNA molecules can be damaged. The activity of the epoxide hydrolase can be reduced as the age progresses. The chs such as heterocyclic substances, polycyclic and benzenoid compounds can produce epoxides which that are intermediate. The epoxide hydrolase can damage tpesticides which can act as carcinogens and mutagens. The organic pesticides are frequently inthe agriculture. These substances can produce epoxides. These epoxides can be used as AIs or active ingredients. Some of the examples of this type are endrin and dieldrin. These chemicals can be weak substrates for the enzyme epoxide hydrolase. The epoxide molecules are not stable. This enzyme is needed for the degradation of the chemicals biologically. The insecticides can be oxidized in the microsomes. The microsomal oxidases are abundant in the microsomes and they can oxidize th carbaryl epoxide. The microsomal polysubstrate oxygenases can be linked with the endoplasmic recticulum. The smooth endoplasmic reticulum plays a major role in this process. These enzymes are present in the liver cellof animals. This enzyme can add some oxygen atoms to the substrates. This enzyme is also calledas mixed function oxidase.

The cytochrome P_{450} of the mono-oxygenase system can be inhibited by methylene dioxyphenols. The chemicals such as piperonyl butoxide can inhibit these enzymes. The carbon monoxide and n-octylamine can be used as inhibitors of cytochrome P_{450} monooxygenase system. The hydroxylation process can often yield compounds which are unstable. Oxygen can be linked with the olefinic bond and there will be the formation of epoxides.

Due to these reactions, there will be the formation of the oxirane ring which is considered as a heterocyclic compound with three carbons. The sulphoxides can be oxidized to form sulphones. The hydroxylation of the aliphatic compounds is crucial. The hydroxylation of the rings can be seen in the pesticide metabolisms. The pesticides can comprise of terocyclic ring structures and aromatic structures. The chlorpropham chemical is a pesticide which that is classified as aherbicide. Ddation reactions, there will be the production of the phenolic derivatives. The omega delakylation dealkylation reaon in most of the pesticides. *Rhizoctonia* can oxidize systemic fungicide chloroneb. N-delakylation dealkylation process can be observed in most of the pesticides. Due to this reaction, the amides and amines can be substituted. Hydroxymethylation reaction can be crucial and there will be the elimination of formaldehyde. In the case of thiophoscticides, oxidative desulphuration can be crucial. In the process of sulphoxidation, the sulphides sulfides can be oxidized to the sulphoxide molecules. The sulphoxidation p can be mediated by the FAD monooxygenase system. Demeton-S-methyl is a systemic insecticide which that cbe modified by this process.

The role of the glutathione can be important in the degradation ofthe pesticides. Glutathione can be called as GSH. It is a tripeptide molecule. The residues which can exist in these molecules are glutamic acid, cysteine and glycine. It is important to know the relation between the glutathione S transferase and the degradation of the pesticides. The glutathione molecules are capable of forming conjugated molecules with the pesticides. The transferase enzyme is crucial in the liver cells of mammals. This enzyme can be produced by the plants. The chemicals such as oxirane or tridiphane and triphenyltin can damage these enzymes. Tridiphane can exhibit synergist functions. Addition reaction is crucial in this entire process. Glutathione S epoxide transferases can comprise of epoxide ring. This enzyme is crucial in combating the carcinogenic oxides. Some of the substrates for this enzyme are phenanthrene epoxides and Some of the organochlorine epoxides are heptachlor epoxide, endrin and dieldrin.

The Glutathions Glutathione S aryl transferase can be predominant in the liver and it is also produced in the plants. The reductive, hydrolytic and oxidative changes can be useful in inserting carboxyl and hydroxyl groups. Due to addition of these groups, the molecules can get polarized. There is a possibility that there will be secondary conjugation. The secondary conjugation can be crucial in the endogenouscules. Most of the reactions which involve conjugations can be based on conjugating agent.Most The most important conjugates are the glucuronides. They are observed inthe vertebrates. Glucuronyl transferases can exhibit specificities which can be overlapping in nature. These enzymes are abundant in the microsomes. The herbicides, fungicides and insecticides can be transformed intoronic acid conjugates. These enzymes are present in the liver tissues. Ferbam is a fungicide which that rtly converted into S-glucuronide. The plants and insects can use glucose conjugates. The sulphate sulfate esters can be important in eliminating the foreign materials fertebrates and amphibians. The enzyme sulphotransferase can mediate the production of the sulphate sulfate conjugates. The glycine conjugates can be present in the vertebrates. These conjugates can be formed mostly from the pesticides which are aromatic in nature. Chlorthiamid can be cleaved to form dichlobenil. The glutathionegates can undergo certain secondary changes and get excreted. The are some other compounds which that can be excreted after certain modifications; they are calledaercapturic acid.

8.3 ORGANOPHOSPHORUS BASED MOLECULES

These chemicals were popular during the II World WarII. The Germans has produced several organophosphorus compounds which have insecticidal activities. They also produced tabun and sarin which are considered as toxic nerve gases. The R groups present in these molecules will be occupied by either ethyl or methyl groups. The X portion can be filled with the heterocyclic group, aliphatic group or homocyclic groups. The organophosphorus molecules can work as insecticides and they are better than the organochlorine compoundshe organophosphorus compounds can attack the insects which are small and smooth. They can targetthe aphids effectively.

The aphids can attack the apples and pears. Some of the top fruits can be damaged by these aphids. Some of the chemicals which can control these aphids are phosphamidon, diclorvos and dimethoate. The codling mouth can also be controlled by these chemicals. Soft brown scales and trips can attack the avocados. Malathion can kill these pests. The red spider, thrips and aphids can damage bananas. Malathion can also control these pests. Almost all types of beans can be damaged by the Aphids. Some of the chemicals which kill these aphids arthion, mevinphos, menazon and dimethoate. The looper caterpillar which can damage the beans can be is controlled by dichlorvos. The cereals can be damaged by leatherjackets, Fenitrothion can used to kill these leatherjackets. The citrus species can be damaged by the red scales, mealybug and Mediterranean fruit fly.

Malathion and parathion can kill these pests. Parathion can kill the red scales. Parathion and trichlorphon can kill the stem borer, leaf miner and antestis bug which can infest coffee plants. Aphids, stalk borer and fruit flies can infest maize plants. Some of the chemicals which can kill these pests are trichlorphon, phorate and dementon S methyl. Weevils, thrips, pea moth and Aphids can attack pea plants. Some of the chemicals which can kill these pests are fenitrothion and Azinphos methyl. Aphids can also attack potatoes and they can be controlled by the chemical oxydemeton methyl. Weevils can also infest the grains which are stored. Pirimiphos-methyl, malathion and dichlorvos can kill these weevils. Aphids and some of the viruses can attack the sugar beets. Mevinphos can kill these pests. Leaf miners can attack the sweet potatoes. Trichlorphon can be used to kill leaf miners. Malathion, phosphamidon and dimethoate can kiiners and aphids which can attack tobacco. Mevinphos and malathion can be used to control americal American bollworm and aphids which infectmatoes. Formothion can kill the red sider mites which damage the tomatoes. Aphids and mangold lfy fly can attack several vegetables. FoOrmothion and dimethoate can kill these pests.

The organophosphorus insecticides can be classified based on the chemical compositions. The subgroup 1 compounds can eow rate of chemical stability. They can be less soluble in water and they can be readily hydrolysedhydrolyzed. Most of the insecticides of this group can be classified as contact type insecticides. Tetrachlorvinphos and mevinphos can be classified as subgroup 1 insecticides.

The subgroup 2 insecticides can exhibit higher chemical stability. They exhibit low water solubility. They can soak into the leaves. They are classified as locosystemic and persistent contact compounds. These insecticides can be activated well before it reaches the location. Some of the examples of this type are trichlorphon, fenitrothion, diazinon and malathion. The subgroup 3 can exhibit higher chemical stability. They can enter the plants and can migrate to all parts of the plant body. They can be classified as systemic compounds. Disulfoton, formothion and phorate can be the best -suited examples of this type. The subgroup 4 chemicals can exhibit higher vapor pressures. They can exhibit a low rate of chemical stability. Fumigants can be classified as subgroup 4 chemicals. Dichlorvos is a subgroup 4 chemical. The subgp 5 chemicals can include organophosphorus insecticides. These pesticides can be used to controlthe soil pests. Bromophos, diazinon and phte can be classified as subgroup 5 chemicals. The insecticides can attack the nervous systeme insects and other animals.

As far as the nervous system is concerned, there are 2 aspects which that must be considered while dealing with the pesticides. The nerve impulse must be passed down so that there will be a muscular contraction. The action potential will get generated due to cell-cell transmission. The neuroeffectortransmitters are norepinephrine and acetylcholine. Among these, atleast one chemical can be presenet in the vesicles of presynaptic complex. The pesticides must remove the transmitter by the process of destruction or diffusion or by sorption. The peripheral nerves present in the insects may not have myelin sheaths. This will determine the selectivity of the insecticides. The neurotransmitter acetylcholine can get associated with the enzyme cholinsterase. Two sites present on this enzyme can acommodate acetylcholine. One site can be called ans ester-forming site and other site is called ans anionic site. The ester-forming site can have serine residues. The anionic site will have glutamic acid residues. The carbonyl group which is present on the substrate can produce a positive charge. This is the reason why there will be an electrophilic attack on the first site which is populated with serine. The hydroxyl groups present in the serine residues can be considered as important. This will result in the acetylation process. Due to this process, there will be cleavage of the acetylcholine and hence it gets deactivated. The choline molecule will get freed and they eventually leave the surface of the enzymes. One molecule of these enzymes can destroy 300000 molecules of acetylcholine. The organophosphorus pesticides can exhibit the same structure as the acetylcholine. The organophosphorus chemicals can phosphorylate the enzymes. The acetylcholine molecule can compete with the pesticide carbamate for the enzyme surface sites.

Atropine can be used as an antidote in case of poisoning due to these organophosphs insecticides. This chemical can block the excess production of the acetylcholine. The effect ofthe organophosphorus pesticides can bethree classes. They are acute toxicity, long term and chronic effects. Some of the organophsophrous organophosphorus compounds may not lower the enzyme acetylcholinesterase. Mevinphos can be considered as a contact organophosphate. The persistence of this pesticide is low. Tetraethyl pyrophosphate or TEPP can be used as an organophosphate. It is a water-soluble compound. It has lower stability but the toxicity of this chemical on higher animals is higher. A single drop of this chemical can be dangerous. Tetrachlorvinphos can be used as a pesticide. Its persistence in the environment is low. It can work as an acaricide. We can usethis insecticide on the bush fruits. Dichlorvos can be used as it has higher vapor pressures. Their solubility in water and the stability after hydrolysis is low so that the persistence of these compounds in the environment is for a short period. Some of the chemicals such as Mevinphos can induce systemic action for a short period. This chemical can be classified as organophosphorus insecticides which can exhibit a higher rate of toxicity. It can be dissolved in water. It is also soluble in lipids. It can be present in the environment only for short perThe persistence time of these pesticides can be based on the weather conditions and other factors which that can cleave the pesticides. This chemical can be considered as toxicant especially at the site of action. The reason behind this toxicity is that there is no proper enzyme setup to neutralize this compound. This chemical can be in cis and trans isomers. Mostly the cis form can be used in the commercial production of these pesticides. The cis isomer is more toxic than the trans isomer. It is calculated that the cis isomer is a hundred times more toxic than the trans isomer. In the case of the cis form, there will be carboxymethyl and methyl groups. The cis isomer is more suible to damage than the trans isomer. It is obvious that if the cis formists for a long period, they it may become less toxic. These forms can get evaporated if they persistted in the environment for long periods. The loco-systemic compounds can be considered crucial in the management of the pests as they are powerful and active. These compounds can exhibit low water solubility and they can be dissolved in lipids. The stability of this chemical can differ and they can exist in the environment for a long period.

We can classify this chemical as cct poison which is persistent. As this chemical is lipophilic in nature, this chemical can penetrateinto the cuticle of the leaves even though the leaves are having waxy cuticle. This chemical can migrate to the shorter distances from the site of action. In some cases, this can travel for long distances. This property is important as it can be used to control the pests in the low growing plants such as strawberries. It is to understand that this chemical cannot be sprayed directly on the strawberry plants. It can be absorbed and may not be removed by rain. Its solubility in water is less. It can translocate into other places. They can get transported via plant roots. Iis important to know that this chemical can get activated even before it reaches the site of action.

The nervous system of the insects can get damaged due to this chemical. The activation of this chemical is due to the oxidation process. Due to the oxidation reactions, it can get more polar nature than the actual molecule. Due to the oxidation process, we can expect systemic action and this action cannot be possible with the original molecule. This subgroup accounts for most of the organophosphorus insecticides in the market. This subgroup can be classified into three families. The families are malathion family, parathion family and heterocyclic family. The toxicity in the mammals can vary with the type of pesticide being used. The toxic liquid among these is the parathion compound. Methyl parathion can be classified as less toxic chemical. The chemical fenitrothion can be considered as safer. Malathion can be classified as loco systemic compounds. Its made up of dithiophosphate and it acts as a succinic acid ester. It can be used to regulate the aphids and other insects which can of sucking type. We can control red spider mites, scale insects, small beetles and weevils with the aid of this chemical. Malathion can be considered as weakly active. Malaoxon can be considered as strongly active chemical. athion can be activated to malaoxon due to the action of carboxylesterase. The hydrolysis can result inthe detoxification. The enzyme carboxylesterase can be used by the microbes so that they can be protectd from malathion. This enzyme can remove the ethyl groups from this pesticide. In this process, the thioln S can be used to replace oxtom.

Malaoxon can be used as an inhibitor of cholinesterase. The insects can have oxidative enzymes which that can combat the harmful effects of the insecticides. The insects can break the P-S link present in the malathion molecule. It is important to know that the important defensystems in the insects can activate the insecticide malathion and kill them. It is important to know thatthe vertebrates and insects contain the NADPH oxidase system and esterase. The cleavage of the malathion molecule can be a more complex mechanism. There are some reports stating that the glutathione S transferase eliminates the methyl groups from the methoxy groups of this chemical. The can define the effect of toxicity on the pests. The organophosphorus - based pesticides can be hydrolysed hydrolyzed with the aid of the esterases. Due to the excess use of this insecticide, the pests have developed resistance. The development of the resistance towards this pesticide was reported from several parts of the world. These pesticide- resistant pests have also generated resistance towards other insecticides. This phenomenon in which resistae towards one insecticide can lead to ton of resistance towards other insecticides can be calledas cross resistance. The pests which are resistance resistant to DDT other organochlorine insecticides can also get resistance towards malathion. Townend and Busvine in 1969 has have described that the insects which are resistatowards malathion have got high rates of carase activity. This is the reason why the insects getthe resistance. In the same way,Dayterman and matsumura Matsumura in 1962 have observed the mosquitoes which are resistant to the malathion pesticide.

The parathion family can be classified into fenitrothion, parathion and methyl parathion. The chemical parathion can exist in the environment for more t than the methyl parathion subfamily and fenitrothion subfamily. These chemicals can regulate the growth ofthe pessuch as weevils, sawflies, leaf miners, capsids and aphids. The parathion subfamily can be used to regulatethe pests such as woodlice, nematodes, root-knot, scale insects and mealybugs. The fenitrothion subfamily can be used against grain wels, grain beetles and flour beetles. These chemicals must be sprayed on the walls of the warehouse so that it they may not be mixed with the food materials. There will be some minor variations in the chemical structures among these chemicals and subgroups. These compounds can be oxidized readily with the aid of the monooxygenases in plants, insects and animals. Due to the oxidation process ,there will be the formavatives which comprise of P=O group. These derivatives can be considered as inhibitors of the chlolineaterasescholinesterases. There will be degradation of the subgroups due to the enzymatic reactions. The parathion subfamily can be degraded due to the cleavage of the P-O-Phenyl linkages. Due to this cleavage,t consider it as a hydrolytic reaction. It is just another oxidative reaction. This reaction can be catalysed catalyzed by the NADPH dependent oxidase enzyme. The chemicals such as fenitrothion anrathion can be destroyed by the activity such as cleavage of P-O-CH_3 bonds.

Faskami and Shishido in 1966 has have found that the cleavage mechanism in the mammals is faster than that of in the insect systems. The parathion absorption is quite higher and about 80% of its absorption by the insect system can be fatal for the insects. Sulfur can be removed from the parathion in the lab conditions. The sulfur molnked with the microsomes. The biding is organic in nature and there will be the formation of hydrodisulphideshydrosulphide. The hydrosulphides can be represented by the bond R-S-S-H. Thesanic sulfur can be eliminugh the excretion processes. The excretory substances can be dominated wi thiosulphates and sulphatessulfates. The glutathione can be used in the production of the inorate sulfate molecules. Fenitrothion can be considered as an important insecticide. The insects can metabolise metabolize this chemical at a slower pace and this can kill the insects as it can exhibit chronic react. Post topic application of this chemical there was rapid metabolism observed in the insects. The transfrease transferase enzymes can remove the alkoxy methyl group.

The fenitrothion can produce harmful effects on the larvae of the insects. Due to this chemical, e will be damage to the various stages of the insects. The larvae will be deprived of the food sources for its their development and hence will die prematurely. The influence of the fenping zone. If the moisture is increased 15 times, then the influence of this pesticide will be reduced by 24%. Most of this chemical will be deposited on the pericarp and it may not be available on the surface. This occurs if there is a high rate of moisture.

The important fami among them isthe diazinon family. This family can leave the heterocyclic groups and hence they can be calledasheterocyclic leaving groups. The members of this family can be called as contact poisons and they will be persistent in the environment for long periods. Some of the other chemicals which can be of this type are triazophos, phosalone and chlorpyriphos. The chemical diazinon can be used as a foliage spray. It canbe used to regulate the small insects and other pests which are soft bodied. Temical can be used to regulate the carrot fly and cabbage root fly. There are many members of this group which that can be used to control several pests present in the soils. Laskowski *et al.* in 1982 explained the 4 loco systemic compounds whilst using these chemicals in the soils. The chemical malathion can exne day and about half of its original concentration can get vaporized. The chemicals chlorpyrifos and diazonin diazinon can survive in the environment or at least 60 days. The exact decomposition rates of these chemicals may not be accurate and these rates will be based on several factors. Diazinon can be tolerated by the tobacco budworm called as *Heliothis.* These pests can damage most of the tomato species. This chemical can act mostly on the acetylcholinesterase molecules. This chemical canbe harmful in several animals of a high order. Chlorpyrifos can be classifoad -spectrum insecticide. This pesticide can control the San Jose scale which attackspeaches. This insectidie insecticide can be used to kill the chinch bug in the lawns. It can also control the mosquitoes andoaches. These formulations can be used to regulate the pests present in the soils. There are many factors which that control the activity of this insecticide. Getzin in 1985 has explained some of the factors which influence the activity of this pesticide. The pesticide Azinphos methyl can be used to regulate the winter moths, tortrix and codling which attacks cane fruit, bush fruit and top fruit. Some of the pests which developed resistance to this chemical are the red spider mites. One of the aphicides whican kill the aphids is the Menazon. This chemical can exhibit systemic activity. This chemical must be applied on to the leaves. But reports suggest that this chemical can be absorbed via roots more readily. Most of the heterocyclic compounds can be active when encountered by the mono-oxygenases. There will be the formation of the P=O bonds and hence the derivatives of this chemical can reduce the cholinesterase.

The systemic insecticides can be soluble in the lipids. Their solubility with water is higher than any other chemicals in the group. They can penetrate via cuticular waxes. These insecticides can pass into the xylem and phloem vessels. These chemicals can exhibit minimal stability. They can be present in the environment or in plants for a few weeks post spraying. The systemic insecticides can have several advantages over other chemicals. They can be wiped off due to rain. This chemical can be transported to various parts of the plant. They may not damage the growing tissue as they transport often. These compounds will have minimal contact action. This is the reason why torganisms which are beneficial for plant growth are not affected. The systemic insecticides can be used againstthe insects which are phytophagous insects. They can kill aphids, red spiders and mites. Some of the pests which can be killed are woolly aphids, weevils, leaf miners, codling moth, sawflies and bulb flies. Phorate can be considered as one of the toxic chemicals which can protect the plants from several pests. This chemical can be used to control aphids and capsids. This chemical can lower the roll virus which can attack the potatoes. We can use this type of chemical to control thellow viruses which can attack the sugar beets. These chemicals can be used as granules and they can be applied on to the soils. The granules can be added to the holes drilled for sowing the beans and peas. These granules can control the larvae of the weevil. These chemicals can control the larval growth of the fruit fly and carrot fly. The granules can be active for 6 weeks. The thioester sulfur atoms can get oxidd to produce sulphoxides. These sulphoxides can transform into sulphones. These changes generally take place inthe animals, plants and insects. The sulphones produced can enhance the toxicity in the mammals. When the thioester moiety is removed, there will be dexification of the pesticides. The carbamate family can be important in combating pests. These chemicals compriseof amides which are important.

Dimethoate can be soluble in water and it can exhibit shorterfetimes. The interval between the two applications must be atleast 1 week. This chemical can be used to controlthe pests of crops such as ornamentals, tobacco, bush fruits, top fruit, beet and other legumes. This chemical can regulate beet leaf miners, pea midges and aphids which attack the soils. Carboxylamidases can be important to control the insects. These enzymes can be in higher quantities in the microsomes. Glutathione S transferases can be used to regulate the pests. The phosphamidon can be divided based on the carbamic acid groups. Their solubility in the water can be crucial in determining the toxe chemicals. Dicrotophos can be soluble in water and it is least stable. It can exhate of unstabilityinstability. It be used to control the pests on the soybean and cotton plants. Trichlorphon Trichlorfon cae used to controlthe pests such as root fly larvae, tortrix and leaf miners. We can use this chemical to combatthe resistant lies and cockroaches. They exhibit less toxicity. Methamidophos can be classified as a systemic insecticide which that is toxic in nature and it exhibits water solubility. It exhibits a shorter life span in the environment. Monocrotophos can be classified as a poisonous substance. This chemical can be water soluble. It exhibits minimal persistence in the environment. This chemical can be used to regrcane, tobaanuts and cotton pests.

The organophosphorus compounds can be used as fumigants. The chemical dDichlorvos can be is used a fumigant. This chemical has high vapor pressure which can make it toxic than other chemicals of this group. At In the beginning, these als were used to combat the houseflies which damage the greenhouseops. They were also used to kill pests which that grow in the mushroom farms. This chemical can be poisonous tothe human beings. This chemical can be toxic for atleast one day. It is a best practice to enter the field with enough protection. This chemical has a high rate volatility. Due to this property, it can be us protect the harvest from the pests. It is most effective onthe insects which are free living. It can attach with to the insects before they reach tinterior parts of the plant parts. The cleavage of this chemical can be based on two methods. There is no need of for activating this chemical as seen in for other chemicals or pesticides. The cleavage of this pesticide can be done by hydrolysis or enzyme- mediated process. The enzymatic process involves the removal of dichlorovinyl group which is crucial. There will be the formation of dimethyl phosphate. This chemical can survive in the environment for atleast 12 hours from its application.

Chlorfenvinphos comprises of vinyl group which shares same properties as dichlorovos. It can be used in the form of granules. This chemical can be used to control the pests which attack brassicas. Some of the pests which can be controlled are fruit fly larvae, carrot fly larvae and cabbage root fly larvae. We can use this chemical to control pests in the mushroom farms. Disulfoton and Diazinon can be used as foliar applications. They can be applied in the form of granules. Phorate can be classified as a systemic chemical compound. They can give protection against viruses and aphids. Bromophos can be applied in the form of granules. It can be considered safe as it will be in the form of granules. Chlorpyriphos can be used to kill the microbes present ithe soils. Isofenphos can be considered as a weak organophosphorus compound.

8.4 CARBAMATES

The chemical calledasphysostigimine can be classified as an anti-cholinesterase agent. This chemical has some medical properties. The carbamates can be classified into different subgroups. The subgroup 1 includes aryl N-methylcarbamates. They are the esters of phenolic compounds. The subgroup 2 can include N-dimethylcarbamate esters. These chemicals can have heterocyclic phenolhe subgroup 3 includes oxime derivatives which are from the aldehydes. Adicarb can be considered as an example for of this type. The hydroxyl group can lated. The ring structures can be needed for a successful insecticidal action. The carbamates can slowly hydrolyse hydrolyze in the acidic and neutral pH conditions. In the case of basic pH conhe reaction will be faster. The span of these carbamates will be 10 days. The amidases and esterases can hydrolyse hydrolyze these carbamates. The amidases can be breaking the bonds which are linked with the N atom. The esterases can break the bond which is linked with the carbonyl groups. The hydrolysis process which is based on the esterases can be crucial for the carbamate hydrolysis. The rate of hydrolysis will be based on the organism and the carbamate being used. The carbaryl hydrolysis can be rapid in the case of humans and sheep. In the case of pig and monkeys, it is quite slow. The carbamate based insecticides can be used to control the insects which may not be killed by the organophosphorus compounds. They can kill some of the insects which are susceptible. We can control the insects which are tolerant and resistant to the organophosphorus compounds. Some of the pests which can be controlled by these compounds are wasps, earwigs, cockroaches, scsects, mealy bugs, ants, leaf miners and whiteflies. We can control the pests such as aphids and other pests which that have developed resistance towards the organophosphorus insecticides.ese chemicals can be used to kill the millipedes, leatherjackets and wireworms. The Alfalfa weevil which attackson the Alfalfa crop can be controlled by methomyl carbamate. Tortrix and codling moths which attack apple crop can be controlled by oxamyl, methiocarb and carbaryl compounds. Several pests which attack the banana crop can be controlled by carbaryl compounds. Some of the aphids which are resistant to the organophosphorus chemicals which attack bean crop can be killed by pirimicarb. The root fly of cabbage which is resistant to thrganophosphorus compounds can be regulated by pirimicarb and carbofuran. The fruit flies and mites which attack on the citrus crops can be controlled by methiocarb and oxamyl compounds. American bollworm which attacks the coffee plants can be controlled by carbofuran and carbaryl compounds. The same pest can attotton plants and they can be controlled by methiocarb, carbaryl and aldicarb. The aphids which can attack the hops plants can be controlled by propoxur, methiocarb, aldicarb and methomyl compounds.

Some of the pests such as corn earworm and corn rootworm can infect the sweet and maize corn crops. The chemicals such as metalkamate, methomyl and carbofuran can control these pests. The steam nematode which can attack the onions can be controlled by aldicarb. The pea aphids and pea moth wttacks pea plants can be regulated by pirimicarb, carbofuran and carbaryl compounds. are several pests which that can infect the peanuts. Aldicarb can control these pests. There are some pesthat can infest pineapple crop. Metalkamate can be used to kill these pests. The potatoorms and colarado coloradobeetlettles can attack the potato crops. Primicarb, aldicarb and oxamyl chemicals canre used to control these pests. Water weevil, stem borer and armyworm can attack the rice crop. The chemicals auch such as carbofuan, metalkamate can be used to control these pests. The millipedes and leafminers which are resistant to organophossphorus chemicals may attack the sugar beet crop. Primicarb and carbofuran with aldicarb can kill these pests. The cane borer of the sugar cane crop can be controlled by the chemicals aldicarb, carbofuran and carbaryl. The root-knot nematode and tobaccrm can attack the onions, carrots and vegetables. These pests can be controlled by methiocarb,methomyl and cararyl compounds.

Most of the symptoms due to carbamate poisoning can toxicity of the organophosphorus compounds. We can observe carc arrest, low blood pressure, low blood pressure,spasms and others. The limbs of these pests will be damaged. Ifthe carbamate poisoning is less, we can observe some abnormal behavioral patterns. Atropine ctrol the harmful symptoms exhibited by the carbamates. Carbaryl can be cidered as a carbamate insecticide which that can influence the pests. The hydrolysis process can be important inthe carbamate metabolism. The mono-oxygenases which are present in the microsomes can activate these insecticides. The compounds which are formed as a result of hydrolysis and oxidation can undergo conjugass. The conjugation process can be based on the species. Some of the conjugation products are phosphates, sulphates sulfates and glucosides.

Propoxur can be classified as an important chemical in the caryl family. They come under the carbamate insecticides. This chemical can be soluble in water. It can be toxic tothe mammals. This chemical can be used as a spray on the leaves. This insecticide can be used to control the whiteflies and aphids. The isopropoxy group can produce a speceature which is crucial.

There will be hydrolytic dealkylation. Methiocarb can be used to regulate pesticides which that attack hops, vegetables and cotton. Some of the pests which attack these crops are mites, leafhoppers, codling moth and fruit flies. This chemical canbe used to treat the seeds and repel the birds. This pesticide can exhibit molluscicidal properties.

There are two heterocyclic monomethyl and dimethyl carbamates which are considered important. Some of the important chemicals which can be of importance in controlling these pests apirimicab and carbofuran. The chemical carbofuran can exhibit LD_{50} values of about 11kkg. It can be toxic tothe humans. The dermal toxicity is lower. This chemical can be used in controlling pests which that can attack the coffee, rice and maize plants. This chemical can be absorbed by the plant roots. The carbofuran molecules can kill the nematodes. The higher concentration of this chemical can damage most of the pests present in the soils. The minimal concentration in which this chemical can be used is 50 ppm. If needed we can use higher concentrations. This chemical can persist in the environment for atleast 50 weeks. This chemical can kill the pests if there is alkaline pH. In the case of sandy loams, the microbial ecosystem can be developed and hence there will be ancellent effect of pesticides on the pests. Some of the species which cannot be affected due to this pesticide are the*Penicillium* and*Fusarium*. The chemical carbofuran can be applied to the soils. We can first observe the lag phase in the soils and then there will be a decomposition phase. This chemical can be used as soil sterilant. Due to the addition of these chemicals, there will be variation in the soil microbes. *Achromobacter*can use the chemical carbofuran and this species can use the nitrogen content present in the pesticide. The carbofuran undergoes oxidative metabolism. The oxidative metabolism can occur in the higher animals especially in the liver cells. The resistance to the insects for carbofuran can be linked with the enhanced lic rate. The carbodulfan can be classified as sulphenylated deriive. It can be used late many organisms which that can grow in the soils. We can use this insecticide to combatthe pests such as flea bettlesbeetles, millipedes and wireworms. The root fly pest can also be killed by this chemical. The granules can be used in the form of formulations to treat the soils. We can apply this pesticide in the furrows. The granules can be used to regulate the weevils which can attack the sugar beets. When this pesticide is applied frequently on the soils, then we can see that some of the pests can be killed. can be classified as an insecticidal carbamate. We can expect the translaminar movement of this pesticide post application-. This chemical can exhibit systemic action on the pests. This pesticide can be used in the form of dispersible granules and wettable powders. N-Methylcarbamate derivatives can be considered as crucial in regulag the pests. Some of the chemicals which can be used are oxamyl, methomyl and aldicarb. These chemicals can be poisonous tothe animals and humans. Aldicarb can be classified as oxime derivative. Methomyl and oxamyl compounds can be classified as thiohydroximidates They can be produced from the imines. The first step in the process of metabolizing this pesticide is the hydrolysis step. ALldicarb can be classified as ester like compound. It can be formed between the hydroxyl group and the carboxyl group. The hydroxyl group can be contributed by the oximes and the carboxyl group can be contributed by the methylcarbamic acid. Some of the pests which can be controlled by these chemicals include millipedes, citrus blackfly, mealy bugs, scale insects and leaf miners. Aldicarb can be used to control the nematodes which are classified as phytophagous nematodes. The nematodes can kill the cereals when they survive in the light soils. The nematodes can be used to facilitate the fungal entry and there will be a chance that some of the viral diseases can infect the plants. The endoparasitic nematodes of the barley and wheat can be killed by these chemicals. The nematodes can kill several vegetables and cause economic losses. The span oe aldicarb in the soil can be based on the nature of the soil and the amount of pesticide applied. Bromilow*et al.* in 1986 has have explained the degradation of the aldicarb in the anaerobic soils. There are several interactions between the pesticides and the nematodes which can be complex. The metabolism of this pesticide on the aerobic soils can be the same as in the higher animals. In the case of anaerobic soils, there can be some portion of ferrous ions, this case, the pesticide can get cleaved in a span of a few hours. The pesticide aldicarb can be considered as an inhibitor ofthe cholinesterase. Methomyl can be classified as a systemic and contact insecticide. We can use this pesticide on the leaves. The toxicity of this chemical is Oxamyl can be daded due to aerobic hydrolysis. They cannot be oxic to the pests.

8.5 ORGANOCHLORINES

The organochlorines can beis also called as chlorinated hydrocarbons. This type of chemicals was prepared duringthe World War II. The heaxachlorocyclohexane and DDT can be considered as main stream organochlorines which can be used as an insecticide. The organochlorine insecticide family can be classified into 3 main families. These insecticides are least soluble in water. They exhibit higher lipophilic nature. The organochlorine based insecticides can be considered as waxy solinormal temperatures. Tchemicals can exhibit long term insecticidal activity. DDT can be effective on plant bugs, hoppers, gipsy gypsy moths, European cron corn borers, codling moths, blossom weevils, Lepidopteran members and hymenopteran members. DDT can exhibit medium- range ctivity on weevils, fruit flies, leaf miners, froghoppers and cutworms. This pesticide can exhibit poor activity in the case of sclelrotinisation and if the pests are having a rapid life cycle. They may not affect the organisms which can live in the crevices of the soils. The mites, white flies, root-attacking inseoil pests, boll weevils and aphids can be least affected with DDT.Methoxychlor can act on the flea beetles, bark beetles and colarado coloradobeetles. HCH can be used in the process of seed dressing. It can be used to regulate some of the pests sucs wireworms, leatherjackets, grasshoppers, weevils, crickets and cockroaches. We can use this chemical to sawflies attacking on the apple. The HCH can control the cotton boil boll weevils and house mosquitoes. HCH can also control Tsetse flies by disrupting their larvaes. We can use HCH as a fumigant. It can be used as a stomach poison. The action forwarded by this chemical can be rapid.

Aldrin can be used on the soils. It can be sprayed on the leaves and the span of this chemical in the environment is short. We can use aldrin to control termites, wireworms, white grubs, stalk borers, scale insects, millipedes, fruit flies, mealy bugs, Japanese beetle larvae, cutworms and weevils. Dieldrin can be used on theace soils. We can use this chemical in the seed dressing process. We can use it in the sheep dips, brain baits and root dipping. The sSucking insects and caterpillars can be least affected by this chemical. This chemical can be used against termites, tsetse flies, scorpions, mosquito larvae, locusts, cutworms, crickets and root fly larvae. Endrin can be used to control all sucking and boring type of insects. Endosulfan can be used to control the wasps, tsetse flies, pod midges, lepidopteran larvae, grasshoppers, beetles and ants, Toxaphene can be used to control boll weevils. Chlordane can be used to control the wireworms, termites, soil insects, moth larvae, earwigs, earthworms, chafer larvae, beetle l The chlordane pesticides can effectively control the grasshoppers and aphids.

DDT can be classified as an important pesticide which that can control the pests and its efficiency is more on the spider mites and aphids. Some of the aphids which are in the stationary coloninnot be influenced by the DDT droplets. Some of the aphids which are mobile can be affected by DDT. The treatment with DDT can go awry away if the pests have developed resistance. The mites can become resistant and so there will be no effect of

DDT. Once the temperature of a region gets increased, the effect of DDT can be lowered. DDT can be used as a fumigant and it can persist in the environment for shorter periods. There will be higher vapor pressures which can be beneficial for the microbes which survive in the hot climates. The cyclodiene family can exhibit strong action and their persistence in the environment is for long periods. The pesticide aldrin can have high vapor pressures. It is important to know that the fumigant action is appropriate for most of the soil microbes. The fumigant action can be useful to control the wireworms in the potatoes. The insecticide dieldrin can be considered as an effective insecticide. It controls many pests and insects. Most of the insecticides which are based on organochlorines can be phytotoxic in nature. This phytotoxicity can be high in the case of the cucumber family.

The toxicity of the organochlorine chemi can be considered as neurotoxic. DDT has some insecticidal properties and this was explained in the year 1942. Biggar*et al.* in 1965has have developed a centrifugation technique to explain the properties of DDT. This chemical can be soluble in some of the organic solvents. Some of the solvents are body lipids and natural oils. The solubility of DDT in the olefinic and some of the hydrocarbons can be high. DDT can damage the central nervous system. It has certain neurotoxic effects. Hyperactivity is common in the case of DDT exposure. Some of the symptoms which can be noticed in the case of DDT poisoning are degeneration of liver and kidneys, weakness, anorexia and weight loss. The DDT molecules can act on axonic membranes. Some of the ATPases can be involvecase of DDT poisoning. The ion transfer can be inhibited due to the disruption of the ATPases. The synaptic transmission can be effectaffected by the DDT molecules. Some of the mitochondrial membranes which are crucial in the production of energy can be damaged due tothe DDT poisoning. Mostly the DDT molecules can disrupt the sodium ion channels. DDT molecule may not influence the axon membranes. of the essential activities such as nuclear division, active transport, oxe phosphorylation and photosynthesis can be damaged sue due to the activity of the DDT molecules. There are three crucial pathways which that can be considered important while studying the DDT toxicity. In the first step, there will be dehydrochlorination of DDE. In the second step, there will be reductive chlorination. In the final step, there will be oxidation. After all the three steps, the chemical dicofol will be formed.

DDE can be derived from the DDT and it is present in the biopsy samples. DDTase and DDT dehydrochlorinase can be used in this conversion. Most of the insects can convert DDT into dicofol. This is possible with the microsol oxidation with the aid of NADPH molecules. The chemical methoxychlor can exhibit fewer tendencies than DDT. Dicofol can be consideredas an important chemical which belongs to the DDT fay. It can be used to regulate the growth of mites and ticks. It can be classified as an acaricide. We can use this chemical to controlthe pests such as scrub typhus and red spider mites. Clorbenzilatecan be considered as an important member of the DDT family. In the 1950s it was used as acaricide and it was successfully used then. It was toxic to several pests and insects. Unlike DDT, it can impose minimicity inthe humbeings. The resistance in the insects can be easily developed due to this insecticide.

The hexachlorocyclohexnecan beis also called as BHC or HCH. The molecule lindane can be sparingly soluble in water and it exists as a white solid which is odourless. The insecticidal activity of the lindane molecule can be similar to the DDT molecules. On exposure of insects to this chemical, the insects can suffer from paralysis and there will be an abnormal increase in the respiration rates. Thetion of these chemicals on the nervous system can be similar to the DDT and this chemical can impose higher toxicity. GABA can inhibitthe neurotransmitters. It is more effective on the central nervous system of the vertebrates. The chemicals cyclodiene and lindane can be used to inhibit the neurotransmitters. The mode of action of the cyclodienes and lindane caimilar. The resistance towards cyclodienes can be crucial in provng resistance to the insects. There are atleast 70 metabolites which that are present in the mammals and birds. They can be present inthe plants and insects. Due to the oxidation reactions and enzymatic reactions, there will be the formation of chlorinated phenols, cyclohexene and trichlorinated cyclohexane. Some of the glutathione conjugates have been identified. Some of the chemicals which are present in the insect bodies re trichlorophenols and mercapturic acid.

The chlorinated cyclodiene family can be considered as crucial molecules in the orgorines family. These chemicals are produced by the condensation reactions of hexachlorocyclopentadiene. There will be hydrocarbons which that are cyclic in nature. There will be 5 to 7 C atoms in this molecule. Diels-Alder reaction occurs during this process and there will be involvement of dienophiles and dienes. The chemistry of these insecticides can be complicated in nature. These molecules are not planar. Twill be chair -like configuration. Mirex can be classified as stomach insecticide. It can be used to kill the snails, wireworms and kill ants. This chemical may not be formed due to the Diels-Alder reaction. The pesticides dieldrin and aldrin can be structurally similar. They can be transformed into dieldrin with the help of enzymes which is NADPH dependent. This enzyme is dominant in the microsomes of the liver cells. The chemical endrin can be produced from isodrin through a process called as epoxidation. The olefin molecule can be considered as less polar than other compounds. There will be high vapor pressures. These chemicals can be used as fumigants on the soils and they can control the root flies and wireworms. Dieldrin molecule can be considered crucial in the regulation of the locusts. This chemical cbe used to control several ectoparasites such as lice and blowflies. The main problem with this pesticide is that they can be toxic tothe cattle and sheep. Due to the oxidation of the oxidases present in the microsomes, there will be the production of compounds such as heptachlor, isin and cyclodienes. The epoxides fod can be considered stable. In the substrata just below the mud water, some microbes can metabolise metabolize dieldrin. This occurs inthe anaerobic conditions. In this reaction, we can observe that 2 chlorine atoms are eliminated. This reaction can be at a slower pace. This chemical can be metabolized by several anaerobic bacteria.

Chlordane can be formed due to the mixing of the chlordane and chlorinated dicyclopentadienes. These chemicals can exist in the 2 geometric isomers. There will be 8 chlorine atoms. The cis -isomer can be classified as more toxic. These isomers can be transformed into olefin. This process is called as dehydrogenation. Post this reaction, there will be an epoxidation reaction. Due to this rthere will be the formation of oxychlordane. These molecules can share properties with dieldrin. Endosulfan can be classified as sulphite sulfite and it is a cyclic molecule. It may not be classified as cyclodiene. There will be a bridged ring which is chlorinated. This chemical ile in water. Due to this property, this chemical may not induce systemic reactions. The internal ester in this molecule can be hydrolysed hydrolyzed to form acids and bases. The product which is used in the market is in the form of beta endosulfan, gamma endosulfan and other isomers. Post deposition of thiemical on the the endosulfan can be transformed due to oxidation reactions. The product formed due to the oxidation reactions are is endosulfansulphatesulfate. This substance can induce a high rate of toxicity. This chemical can be produced by some fungi and they canhydrolyse the chemical endosulfan into diol. It can be used as insecticide and acaricide. It can exhibit a high rate of toxicity. This chemical can damage fish in the freshwater environment. This chemical can control the gall mites. It also controls the mites which can attack blackberry and stry.

The insecticiwhich are classified as chlorinated cyclodienebe coered as toxic. These chemicals are banned in mostof the countries due to its their toxic nature. There are many properties which that are shared by the lindane and cyclodienes. These chemicals can induce neurotoxic effects. The primary target of this chemical is the central nervous system. The insects will fail to fly when exposed to the poisoning with this chemical. The activity of this chemical can be enhanced with an increase in temperature. The cyclodienes can be classified as non-axonic poisons. Most of the organochlorine compounds can persist in the environment for some years. This chemical can stay in the soils for longer periods. This chemical can work better if applied to soils and it is the best option if applied on the soils rich in organic matter. Excess volatility can be the main cause of the limited persistence of lindane. The organochlorine compounds can be used to kill bacteria and soil fungi even at normal concentrations. These chemicals can produce much disturbance in the equilibrium. DDT, when applied to soils can reduce the insect family named collembolan. These insects exhibit a decline in their growth rates but eventually, they can increase due to enhanced resistance. There may be some problems associated with this chemical which are yet to be seen as indirect signs of damage are visible. This chemical can be increased in terms of concentration in some of the species and we call this rocess as biomagnifications. The ratio between thentration in the outer environment and inside the body environment can be called as biomagnifications factor. In the case of plants which that can go for high rates of transpiration can take in these chemicals via roots. These chemicals can get stored in different plant parts which are considered edible. The gamma HCH can be absorbed readily. This chemical can cause disturbed nuclear cleavage in the meristematic tissues. The biomagnification can beapparent in the crop roots, earthworms and slugs. We can observe a high rate of biomagnifications in the earthworms. The biomagnifications factor can be measure at 73 in the case of earthworms. Slugs and crop roots can exhibit 0.1 and 4 biomagnification factors respectively. This biomagnifcations factors can be seen in the soils primarily. The process of biomagnification occurs in water too. Fish can have a biomagnifications factor of 829300 and clams have 7000. The pesticides can exist as residues in water. They form due to the run-off from the agricultural lands, animal dips and mostly in the effluents generated from the factories. The contamination due to the organochlorines can be negligible in the drainage water. The main reason behind it is that the soils exhibit more affinity towards the organochlorines.

Matsumura in 1975 has explained the effects of DDT in chronic modese exposed to DDT there will be a disturbance in their behavioral patterns. When the higher animals are exposed to DDT, the organichlorine organochlorine bacides, Phenobarbital and polychlorinated biphenyls can cause damage to the liver tissues. The enzymes present in the liver can metabolise metaboize these micals. There will be enlargement of the liver. These compounds can contaminate eggs, milk and meat and this is considered as threat tothe humans. The food chain effect can be consideredmportant as these chemicals can exhibit a high rate of lipid solubility and high stability. The fish- eating birds can be considered as a the best exao explain the food chain effect. These birds can be harmed due to organochlorine compounds.

8.6 PYRETHROIDS

There are 6 compounds which that can be extracted from the *Chrysanthemum cinerariaefolium*. We can get these chemicals from the dried inflorescence of these plants. We can observe the knock-down effect on the insects' post -application ofthese chemicals. These chemicals can be used against mosquitoes and flies. Pyrethroids are considered as low toxic chemicals. The pyrethrium plants which can grow luxurioin the volcanic regions can produce large amounts of pyrethrin. The flowers were harvested by hand and the pyrethrin chemical can be used extracted from the dry inflorescence. The flowers which are dried can be used for the production of this chemical. In the pyrethrum flowers, there are 4 important ingredieney are Pyrethrin I, Pyrethrin II, Cinerins I and Cienrins II. These chemicals can exist as esters. There are basically acids with a three three-carbon rings. Some of the chemicals which can be available in trace quantities are Jasmolins I and Jasmolins II. The main difference between these two chemicals is that the alcohol groups are saturated. The type I compounds are dominated by chrysanthemic acid. The type II chemical can be dominated by pyrethric acid.

These acids can be linked by the alcohols. Naturally forming pyrethroids can exist in both cis and trans forms. Each compound can exist as an enantiomer. The trans isomer can be dextrorotatory in nature. The trans isomers can be considered as toxic molecules than the other forms. We can observe double bonds and hence one can expect geometrical isomerism. The esters of these molecules can behey are soluble in alcohols and petroleum. Most of these esters are unstable. In the presence of water, these chemicals can be hydrolysed hydrolyzed and so they may not be active against the pests. They fail to act as stomach poisons. Most of the processes involved in the metabolism of these pyrethrins can be based on the hydrolysis reactions. As a result of oxidation, pyrethrin I will be produced.

The pyrethroids which formed naturally can be used both indoors and outdoors. They can be useful in sustaining animal health, medicine and hygiene. These chemicacan be used to regulate fleas and lice. We can control mosquitoes and houseflies with these chemicals. These chemicals are less toxic tothe humans. We can observe minimal toxicity to the warm blooded animals. These chemicals can be destroyed due to cooking or uneven heating. Pyrethroids can be observed evenn the fingers of children.Ruzo in 1983 has explained that these chemicals can be degraded due to the sunlight and hence their applicationin outdoors is questionable. It is perfectly safe for higher animals and some of the synthetic pyrethroids can kill the fishes. These chemicals can kill pests only and the crop of our interest can be saved. We can say that these chemicals are seleive in nature. It is important to know that these chemicals can produce damage to the aquatic environment. The synthetic pyreth analogue was produced and named as allethrin. The properties of allethrin were explained by Schechter in 1949. This chemical is merely an alcohol connected with the chrysanthemic acid. This chemical has produced a weak knock down effect. We can use synergists to enhance the action of the alletrins.

Another synthetic pyrethroid produced is bioresmethrin. This chemical was produced in the year 1966. It can instantly kill the houseflies. They can exist as natural esters. Permethrin can be classified as a synthetic pyrethroid which is of much importance. This chemical exhibits minimal knock-down potential. It can be toxic to many insects. This chemical can be stable even in the sunlight. Among all the synthetic pyrethroids produced, Fenvalerate can be considered important. It has an alpha cyano group. The structure of this chemical is complex enough to make it stable under the sunlight. Some of the moleculesch share the same properties are deltamethrin and cypermethrin. These molecules can also have alpha cyano groups. Gammon *et al* in 1981 has have proposed two pyrethroid classes. They are named as type I pyrethroid and type Ii pyrethroids. The former type does not have alpha cyano group and the latter can induce writhing syndrome.

Permethrin can be considered as a non-systemic pyrethroid. We can use this chemical in controlling pests as it can persist in the environment for long periods. This chemical can kill most of the insects which are phytophagous in nature. These compounds can kill coleopteran and lepidopteran members. We can use this chemical in veterinary medicine and domestic hygiene. We can also use this chemical to control the mosquitoes. Mostly these pesticides can be used in cis and trans forms. It exhibits lower water solubility and less volatility. The initial modifications of this chemical occurs due to oxidation and hydrolysis. The cis permethrin can be metabolized less rapidly than the trans permethrin. This pesticide can attack at specific sites of the hosts. These chemicals an be produced as secondary metabolites.

When the temperatures get decreased, these chemicals can be more toxic to the insects. That means, the effectiveness will be maximum in case ofr days. *Heliothisvirescens* can be considered as sensitive to this pyrethroid. Cypermethrin can be considered as an important compound which that does not have an alpha cyano group.

This chemical can exist in both cis and trans forms. Most of the birds can be resistant to this chemical. This chemical has a wider biological spectrum. It can be used against the sucking and biting insects. This chemical can take hours to kill the vectors. Cypermethrin can be degraded by some species. Fenvalerate can be used as an important pyrethroid. It comprises of alpha cyano group. It can be classified as an insecticide with a broad spectrum. It can be used against insects. Some of the pests which can be controlled by this pyrethroid are pink bollworm, leaf rollers, tobacco budworm and boll weevil. Deltamethrin can be called as decamethrin. This substance can be the same as cypermethrin. It has a cyclopropane ring. We can also observe an alpha cyano group in this molecule. It has 2 Br atoms in the side chains. It can be classified as a powerful insecticide. Its solubility in water is less and mostly it can be soluble in organic solvents. It exhibits low toxicity. It can be classified as an insecticide with a broad -spectrum. It can kill harmful species such as *Apantelesplutellae.* Hydrolysis can be considered as the first step in the process of metabolism. Delrin can exist as a cis isomer. It can be transformed into trans isomer slowly.

Tefluthrin can be considered as an important pesticide which that was developed by Jealott's Hill in the UK.

This chemical can protect the crop from dipteran and lepidopteran insects. There will be 4 fluorine atoms present in this molecule. It has less stability and volatility. We can expect residuals of this pesticide in the soils even after weeks. We can use this chemical in the form of granules. We can use it on the seeds and furrows. There is less toxicity observed with these chemicals. The half-life of this chemical can be for several months. We can expect a knock down effect on the insects imposed by these pyrethroids. DDT can impose several physiological and behavioral effects on the pests as we know. There are several side effects due to the usage of these chemicals.

8.7 SIDE EFFECTS

DDT can produce convulsions, prostration and ataxia with coordination loss in the pests. There are several side effects caused tothe humans when the pyrethroids are used as pesticides. Type I and II pyrethroids can be toxic in nature. Deltamethrin can produce depolarization of the membranes. The sensory nerve fibrils can be most sensitive to these pyrethroids. There will be an induction of discharges in the nervous tissues. GABA receptors are present in the synaptic membranes. They can be linked with the chloride channels. The GABA activation can enhance the chloride ion permeability. The enzymes, channels and receptors can show complexity. It is important to follow caution as they can induce paralysis. Due to this, the insects will die. There will be enhanced convulsions due to the activity of the pyrethroids. Some of these pesticides can exhibit a positive temperature coefficient.

8.8 OTHER INSECTICIDES

The acylureas or benzoylphenylure exhibit some features which can be stated as unusual. The acylurea can be used inthe integrated pest management. There are many insects which that can be resistant to contact action offered by the pesticide. If the plants resist the intake of the pesticide via cuticle, we can say that the pesticide ismful. Diflubenzuron can control the cotton leafworm. Diflubenzuron was prepared in 1973. These molecules can be used to kill the growth ofdipteraldipteran, coleopteran and Lepidopteran. Chlorfluaron can regulate the lepidopteran pests such as *S.littoralis*and *Heliothis spp.* Which infest on cotton plants. These chemicals can be consideredas toxic. This chemicale used to control soybean looper, gypsy moth and codling moth. Flufenoxuron canbe classified as acylurea. It exhibits acaricidal properties which that are strong. It can persist oneaves and the activity on the pests will be rapid. This chemical can be insoluble in water. It can be e in xylene and acetone. It can aatck attack on the larval stages of mites and insects. Codling moth can be considered as an important pest which that can damage the walnuts. We can use azinophos to control the aphids population. These chemicals are highly toxic in nature. We can control aphid parasites with this chemicat of these chemicals can act n the chitin production in the larval stages of the insects. Most of the insects have complicated life cycles which that involve eggs, larvae, moulting and other stages. There will be a pupation and at last, we can observe the imago stage. There will be a production of the exoskeleton in the process of ecdysis. There will be synthesis and deposition of chitin molecules. Chitin can be considered as a crucial component of the cuticle layer. Chitin can provide mechath and flexibility to the cuticle. It can be regulated many substances which can enhance growth. The chitin can associate with the sclerotised sclerotized protein and wax. This association can helpthe insects. It is crucial to know the dynamics of chitin in the growth stages itself so that the pesticides sprayed will be effective. Chitin is made up of a polysaccharide substance. Monosaccharides can join together to produc polysaccharides. Most of the enzymatic reactions can control the activy of these pesticides. The cuticular components canbe produced by the moulting hormone. The cuticle of larvae can be based on the steroid calledasecdysteroid which is classified as a juvenile hormone. The metabolite which is active can be a strong toxicant in some cases. The acylcan be used to damage the production of chitin molecules. Once these insecticides act on the formation of chitin, there will be other steps which that can malfunction so that other molecules may not be affected.

The diflubenzuron can be degraded by microbes which grow in the soils. Some of the microbes which can metabolize or degrade diflubenzuron are *Rhodotorula, Penicillium, Cephalosporium* and*Fusarium* species. Some of the metabolic products of this chemical are chloroaniline, chlorophenylurea and difluorobenzoic acid. Hydrolysis can take place in the metabolism of this chemical. This pesticide can enter into the pupae of the insects and it gets rapidly degraded. Chlorfluazuron can be considered as more toxic than diflubenzuro The chemicals named as sesame and piperonyl butoxide can be considered as inhibitors. These chemicals can disrupt the mono-oxygenase tems which is are a common system in many species. The oxidative metabolism can be damaged.

Insect growth regulators can be effective inhibitors ofthe insect populations. Out of all the growth hormones involved in the life cycle of insect, methoprene and juvenile hormone can be considered crucial. The steps in the insect life cycle comprise of metamorphosis and ecdysis. These stps involve several growth regulators. The insect growth regulators can be designated as IGRs. They can be artificial or natural. They are in some cases, can exhibit antagonistic properties. The acylureas can be used to damage the chitin synthesis. The primary mode of action will be on the chitin synthetase. Some of the chemicals which can be similar to these IGRs are the juvenile hormones. These hormones can be crucial in the development of g, instar stages, pupa and imago stages. Some of the important juvenile hormones are hydropene and methoprene. The growth regulators, pheromone analogues and juvenoids can be crucial in exng the insect growth hormones and their relation with the pests. Formamidine insecticides and acaricides can be considered as important chemicals which that can be used to control the pests. There are two members which are considered important. They are amitraz and chlordimeform. The formamidines can be used to disrupt the normal behavioral patterns in the insects. The acarines can be used to control the phytophagous mites. The general appetite of the organism can be disturbed. These organisms can get killed due to limitations in the food absorption patterns. The spider mites can get killed and fall on thethreads. *Bemisiatabaci* or the adult whitefly can infect the cotton plants. When chlordimeform is sprayed on the leaves, these adult whiteflies may die. Some of the pests such as rice borers can die as the swimm capability of these pests will be impaired. The behavioral patterns can get disrupted and so as the reproduction styles. The sex hormones needed forthe successful reproduction can get ed or inactivated. The moths which are affected by these insecticides can be able to fly in the broad day light and often rests in the night time. Most In most of the cases, it was observed that the sexes may not be demarcated post-mating. The number of eggs produced can be decreased due to the activity of this chemical. The amidines canbe used to regulate the lepidopteran and hemipteran insects. Two species can be affected due to the action of these pesticides. They are *Spodoptera* and *Heliothis.* Formamidines can be used to control the insects. Some of the insects have produced resistance towards these insecticides. This inticide can be used to regulate the rust mites and spider mites if applied with theofentezine. The formamidines can be used to control the mites intheintegrated pest management. The neurohormones and neurotransmitters can involve inthe neural activity. The neurohormones can be used to act on particular sites of the nervous system. The neurohormones can be used as a neuromodulator. This hormone can be used to transmit the information via a synapse. Noradrenaline and adrenaline can be used to control several invertebrates. This chemical can be closely linked with the octopamine. It has several of the catecholamines. Chlimeform can be activated on the nervous system of several insects. The chlordimeform can undergo N demethylation. Demethyl Dimethyl chlordimeform canbe act immediately on the pests. The nornicotine and anabasine can have similar properties of nicotine. These chemicalan act as viscous liquids. On exposure to light, the nicotine molecule can get da nicotine tannate can be insoluble in water. It can work as like a poison in the stomach. The nornicotine and nicotine can be present in the Solanacean Solanaceae members. The genus which has a high rate of nicotine belongs to *Nicotiana.* The nicotine can be produced naturally. The chemical nature of the nicotine can exhibit the laevorotatory mode of polarity. This chemical canss toxicity on the insects. Nicotine can be bind to the acetylcholine ors of the nervous system.

Rotenoids are a group of insecticides which are produced from the plants. This insecticide can regulate ectoparasites which that survive on the animals. Most of these insecticides are present in the roots of the legume plants. *Derris spp.* and *Lonchocarpus*plant specin produce this insecticide. Rotenoids can be used to paralyze fish and they can stay afloat on the surface. Rotenoids can be extracted from *Derris* palntplant. They can be produced in excess quantities from the roots by adding clay diluents in the mixture. Rotenone can be considered a laevorotatory molecule. The formamidines can react with the enzyme adenylate cyclase. This enzyme can be sensitive to the octopamine. It can be used in roduction of cAMP or cyclic AMP. Due to the cumulative effect of these chemicals, we can observe abnormal behavioral patterns. If the formamidines were is used in higher centrations, we can observe anesthetic effects. In this case, glutamate can be a neurotraitter and it will be affected. Cotinine can be observed inthe vertebrates and insects as a result of metabolism. This chemical can be produced due tothe hydroxylation reactions. There will be a dehydrogenation process. The rotenone can be used to suppress the electron transport mechanism in mitochondria. Due to the disruption of the electron transport chn the mitochondri conduction property of the nerve cells will be damagemainly targets flavin mononucleotides.

Acaricides can be consideredas toxic moleculwhich that can kill mites. The mites can be important pests which that can attack several vegetables and fruit trees. Some of the red mites can attackthe greenhouse crops. Most of the white mites can feed on the grains if stored for long periods. Azobenzene can be used in the greenhouses to kill the pests. We cannot use this chemical outdoors. We can add phenyl rings to this molecule so that the toxicity can be enhanced.Tetadifon can be considered as a tetrachlorinated derivative. Chlorfenson can produce severe phytotoxic effects. Due to this, there will be scorching of the leaves and the epidermis of the plants wilet teared off. This insecticide can be used to influenthe nymphal stages and eggs. This ide has lower toxicity in mammals. The LD_{50} value inthe humans can be 2gm/Kg. Another successful acaricide isthechlorbenside. This molecule has a sulphur sulfur atom with bridges. This compound can get oxidized to form sulphonates and sulphoxides. This insecticide can persist for long periods. This pesticide can penetrate through leaves. Cycloprate can be used to kill the phytophagous mites. This pesticide can exist as hexadecyl ester. It can work as ovicidal and acaricidal molecules. It can kill the mites and insects in their egg and larva stages. Cycloprate can be used to break metabolism.

This occurs due to the malformation of the fatty acids in the mitochondria and its irregular transfer. The carnitine molecule can be classified as a carrier molecule for the fatty acids. This transport molecule can be blocked by the chemical cycloprate. Due to these reactions, this chemical can be classified as toxic. Due to the action of this chemical, there will be less proportion of carnitine molecule in the adult mites and eggs. Dicofol can be classified as a DDT derivative. Due to the hydroxylation reactions, the toxic nature of this chemical can be reduced. Fungicides can share some properties with acaricides. These chemicals are dinitrocyclohexylphenol, binapacryl and dinocap. These chemicals can regulate powdery mildews. Clofentezine can be used to regulate the mites on the fruits. This chemical can induce ovicidal action on spider mites. This chemical can be used to kill the pests in the greenhouses and outdoors. This chemical can be effective at 16^0C. Cyhexatin can be used to control phytophagous mites. This chemical can be effective on the mites which infect raspberries, plums and apples.

The nematicides can regulate the phytophagous nematodes which can attack plants especially the plant roots. The nematodes can enhance the viral infections, stunted plant growth and lesser yields. Most of the nematicides can be used in the form of fumigants. These compounds can be applied into the soil via sprinkling or injection. We can follow the rotary type of cultivation. These chemicals can be used in the form of volatile liquids. Mostly they can enter into the crevices present in the soils. Some of the chemicals can exist as solids which are classified as non -volatile solidhe D-D mixture and dichloropropene can be used as nematicides and they regulate the root-knot and potato root nematodes. We can apply these chehe a depth of 30 cm from the surface. Some of the soil sterilants which can be used to regulate the pests are metham sodium, chloropicrin, carbon disulphide disulfide and formaldehyde. These chemicals can control nematodes, fungi and insects. Metham sodium can be applied to the soil. This chemical can be transformed into methyl isothiocyanate.

9

Pesticidal Resistance

CONTENTS

9.1 INTRODUCTION

The crops such as cotton, soybean and corn were used in the form of genetically modified crops. These crops can be treated as herbicide- resistant crops. These crops can be resistant te glyphosate which is considered as a herbicide with a broad spectrum. The glyphosate- resistant crops gained popularitnthe recent times. The herbicide- resistant crops can be called as HR crops. They can be classified as glyphosate -resistant crops as glyphosate can be used to control several weeds. These resistant varieties have helped many farmers all over the world to get benefitted. There are two tras which that can code for EPSPS or 5-enolpyruvylshikimate-3-phosphate synthahe gene cp4 esps gene can be extracted from the soil bacteria named Agrobacterium tumefaciens. Themers has have adapted for the GR crops so that the crops will not get damaged due to the herbicide activity. The weeds which damage the crops will die. There are many hees which that can act as glyphosate. The herbicides can be used to control the weeds. These herbicides can be used to control the weeds which infect the cotton, soybean and corn. The chemical glyphosate can be classified as broad -spectrum herbicide. This chemical can be non- selective in nature. They can applied on to the leaves. The residual activity in the soil is negligible. This herbicide can control broadleaf weeds, sedge, herbaceous grass and other woody plants. Many of the formulations of glyphosate can be used in combating the menace produced by the weeds. The salts which can be used to regulate the weeds are isopropylamine and monopotassium.

The adjuvants can be useful in controlling the weeds. The amount of herbicide added and the timing of its application can be considered to determine the efficacy of the herbicides. The herbicide glyphosate can be a competitive molecule over PEP or phosphoenolpyruvate. This competition between these molecules can occur in the chloroplasts. Due to this enzyme, there will be a disruption of the shikimate pathway. The products obtained from this pathway can include amino acids such as phenylalanine, tyrosine and tryptophan. These amino acids can be considered as aromatic amino acids. They are considered as plant metabolic products. here are some fale reactions which that occur in the plant systems. Some of the characteristics are higher mammalian toxicity, flexibility, soil absorption and low cost. The herbicide- resistant weed management can be tackled with several tactics. One of the ortant tactics isthe herbicide rotation. The benefit of the herbicide rotation is the regulation of the herbicide- resistant weeds and lowered selection pressur Some of the risksof associated with this process are less weed spectrum and lesser phytotoxicity. The impact estimated from this method can be excellent. The process of using herbicide mixtures can be considered as a good tactic. Some of the benefits associated with this method are broad weed spectrum, enhanced control and lowered selection pressure. Some of the risks associated with this method are the phytotoxicity, higher costs and poor weed activity. The herbicide mixtures canit high impact. Otother technique which that can be used is the variable application and timing. Some of the benefits are effective herbicide use and the best control of the HR species. Some of the risks associated with this method are lack of residual activity. The impact can be good with this method. Adjusted herbicide rates can be considered as an important tactic used in combating the herbicidal resistant plantsome of the benefits is are excellent control of the target species. The risks associated with this tactic are the evolution of the polygenic resistance, enhanced selection pressure.

The impact can be poor. Precision in the application of the herbicides can be considered as onehe crucial tactics,.The use of this tactic is that there will be lowered selection pressure and herbicidal use. Some of the risks involved with this tactic are enhanced control variability and enhanced application costs. The impact can be poor. In the case of primary tillage tactic, there will be lowered seed bank of the weed species, decreased selection pressures and enhanced efficacy. Some of the risks involved in this tactic are more time involved and enhanced soil erosion with incd costs. The impact with of this tactic can be good. The tactic involving mechanical weed control can be useful. With this tactic, there will be lowered selection pressure and it is inexpensive. Some of the risks associated with this tactic are more time requirement, crop injury, more skill ne and there may be need of for addit tactics. The impact with of this tactic can be poor. The crop rotation and selection tactic can be beneficial. Some of the uses are lowered selection pressure and variation in the agro-ecosystems. Some of the risks involved with this tactic are limited research base, an insufficient adaptation of the crop to the crop rotation tactic, herbicides and alternative rotation crop. This tactic can be good. Adjusted planning times can be the best tactic. The benefits of this tactic involve lowered selection pressures and efficacy over the weeds. Some of the risks involved are increased rotatio diversity and the need of for alternative strategies. The impact due totic can be pousted The adjusted seeding seedling rate can be considered as a crucial tactic. Some of the benefits of this tactic can be lowered selection pressure and enhanced ability of the crop while competing with the weeds. Some of the risks involved are lowered yields, enhanced intraspecific competition, increased pests and seed costs. The impact of this tactic can be good. The configuration of planting can be used to enhance the ability to compete and there will be lowered selection pressure. Some of the risks associated with this tactic are limitatin in the usage of equipme and less availability ofthe mechanical measures. The impact of this tactic can be good. The usage of intercrop systems, mulches and cover crops can be considered as an efficient tactic. The benefit of this tactic is allelopathy, enhanced system diversity and lowered selection pressure. There will be enhanced competitiveness among the crops and the weeds.

The insects can become resistant if the species can tolerate the pesticide concentratihere are some insects which that can get resistant more quickly than any other insects. The insects can become easily resistant to organochlorine insecticides. The selection pressure of the insecticides can help the insects to retain their resistance. The persistent pesticides can produce unnatural selection and due to this several pests can die on exposure. The breeding process can be faster if there is no predator. Most of the organisms can produce offspring which are resistant to a specific pesticide. The resistance developed by the pests induce serious losses inthe humans. Due to these resistant varieties, the environment in which they grow can become vulnerable. The factors which can influence the resistance in the insectscan be based on the behavioural, biochemical, physiological and morpological nature. The behavioural changes can decrease the frequency of the insects which feed on a specific toxicant. The insects which are flying can pick minimal toxicant than those feeding at sedentary state. Filber aphids have developed avoidance towards toxicants.

The morphological variation can yield less amount of toxicant in the insect body. If we alter the structure of the cuticle then we can observe the morphological variations. Some of the physiological changes can be based on the pattern of absorption and secretion of these insecticides. The biochemical changes can be divided into two classes. There will be a deviated route of metabolism and loss of cell's susceptibility towards a specific insecticide. Insects or pests can gain resistance from one pesticide and the same resistance can be extended to other pesticides. This process is called as cross-resistance. Another process which can be similar to the cross -resistance is the conditioning. The process of resistance can be incurred by a barrier or mechanism which can drastically lower the quantity of the pesticide to reach its target. The barrier and its structure can be crucial in defining the resistancs a specific pesticide. Barrier A barrier can be considered crucial in explaining the resistance. The changes in the cuticle of the insects can be considered important as that will decide the quantum of resistance towards a specific insecticide. The change in the absorption rates is important to understand the resistance incurred in certain pests. If the penetration of the insecticides is regulated or lowered in the cuticle, then there will be increased resistance in the pests. The time of toxicant and the resident times can be important in deciding the resistance. Due to the presence of the mutant gene in some of the houseflies, the pesticides dieldrin and DDT may not kill all the insects as they would have developed resistance.

The excretion process of the pests can be crucial in determining the resistance of the pests. The physiological condition of the insect can decide its resistance towards insecticide. In the case of organophosphorus compounds, the resistance in the pests can be attributed to the activity oydrolases. The resistant starin strain can produce some molecules which can damage the insecticide and it can modify enzymes so that they can disrupt the insecticides. Hydrolases can be crucial in the metabolism of pesticides. These enzymes can be used to produce resistance in the insects. Some of the compounds which get affected due to hydrolysis are carboxylesterases, aliesterases and phosphodiesterases.

9.2 HERBICIDAL RESISTANCE

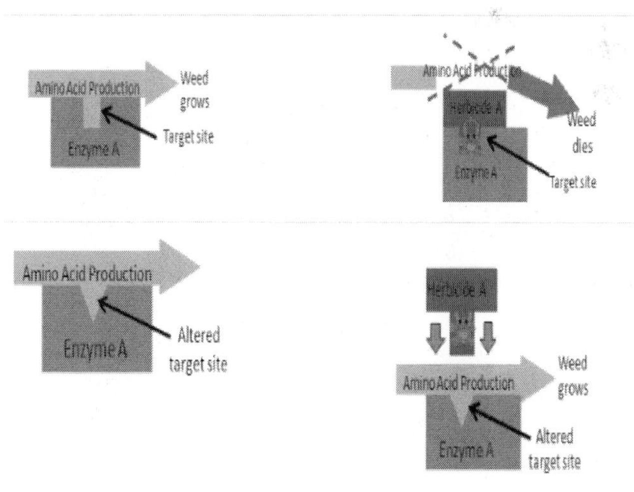

Figure 9.1. Altered Target site.

Source: https://pesticidestewardship.org/resistance/herbicide-resistance/mechanisms-of-herbicide-resistance/

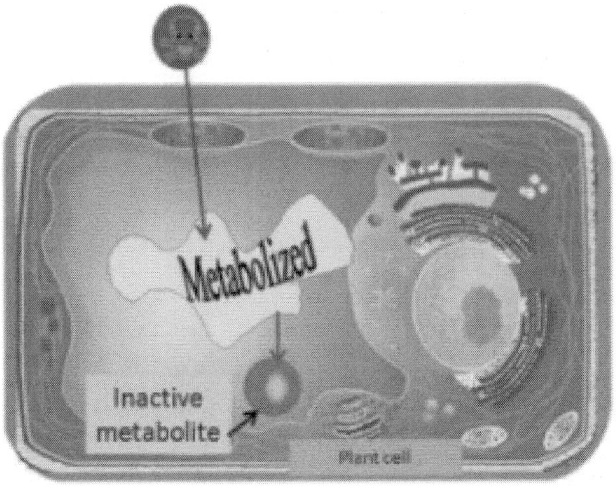

Figure 9.2. Enhanced metabolism.

Source: https://pesticidestewardship.org/resistance/herbicide-resistance/mechanisms-of-herbicide-resistance/

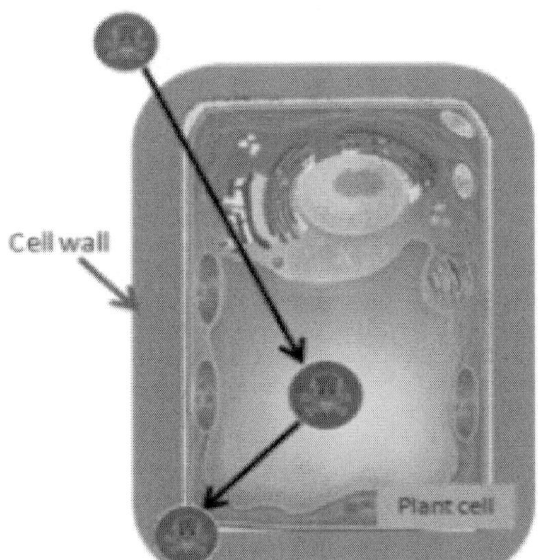

Figure 9.3. Sequestration.

Source: https://pesticidestewardship.org/resistance/herbicide-resistance/mechanisms-of-herbicide-resistance/

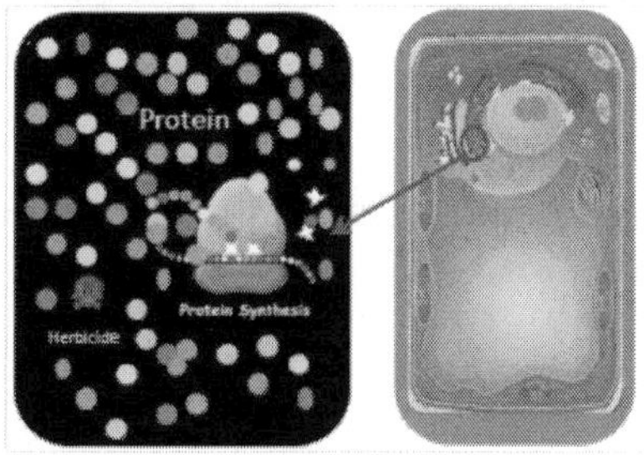

Figure 9.4. Overexpression of target protein.

Source: https://pesticidestewardship.org/resistance/herbicide-resistance/mechanisms-of-herbicide-resistance/

If the receptor for the herbicide is altered then the herbicide may not affect the growth of that weed. This is explained in figure 9.1. Due to the enhanced metabolism of the herbicides in the plant cells, the herbicide may not be useful to control the weeds. This process or situation is shown in figure 9.2. In some plants, the herbicides may not be able to migrate iide the cell. This process is calledas sequestration. This is shown in figure 9.3. There can be overexpression of the target protein and hence the herbicides may not be able to work on the weeds. This is shown in figure 9.4.

9.3 INSECTICIDAL RESISTANCE

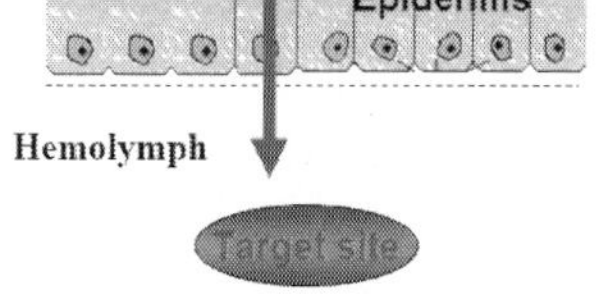

Figure 9.5. Reduced penetration.

Source: https://cals.arizona.edu/classes/ento415/LECTURES/ENTO415_Resistance.pdf

There will be modified structure and composition of the epidermis and hence the rate of penetration will be reduced. This is represented in figure 9.5.

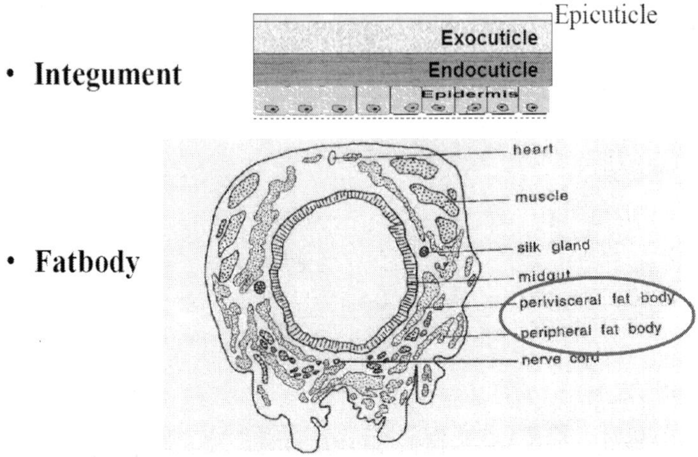

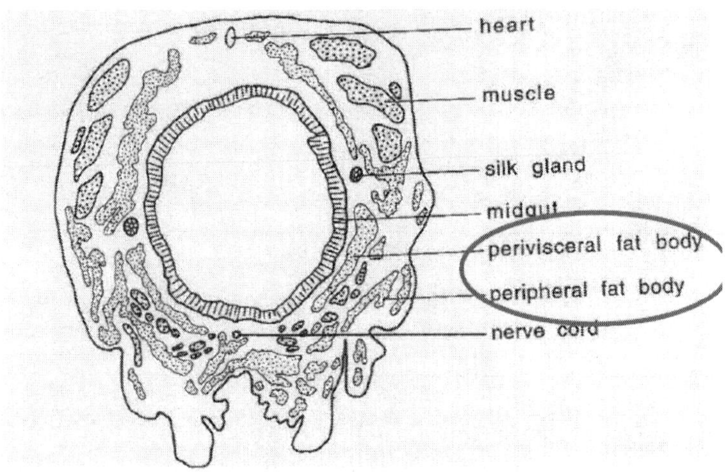

Figure 9.6. Enhanced sequestration.

Source: https://cals.arizona.edu/classes/ento415/LECTURES/ENTO415_Resistance.pdf

Soe of the resistance mechanisms involve Bbehavioral and pentration resistance. Figure 9.6 shows enahanced sequestration. There will be direct excretion. The direction excretion mode of resistanceis shown in figure 9.7.

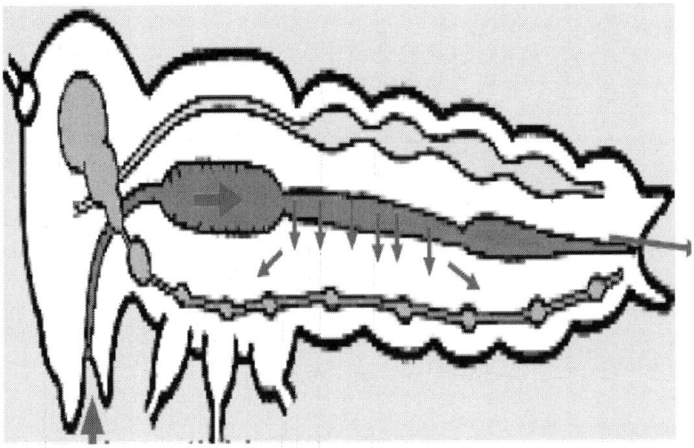

anct excretion (Tobacco hornworms).

Source: Zhu, Fang & Cui, Yingjun& Walsh, Douglas & Lavine, Laura. (2014). Application of RNAi ses/ento415/LECTURES/ENTO415_Resistance.pdf

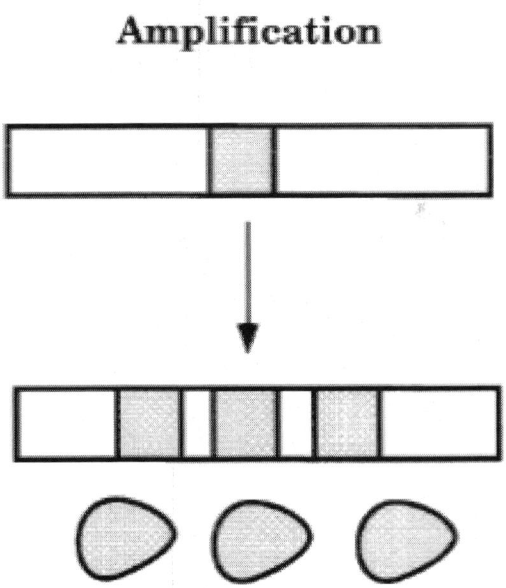

Figure 9.7. Molecular mechanisms of resistance (Amplification).

Source: https://cals.arizona.edu/classes/ento415/LECTURES/ENTO415_Resistance.pdf

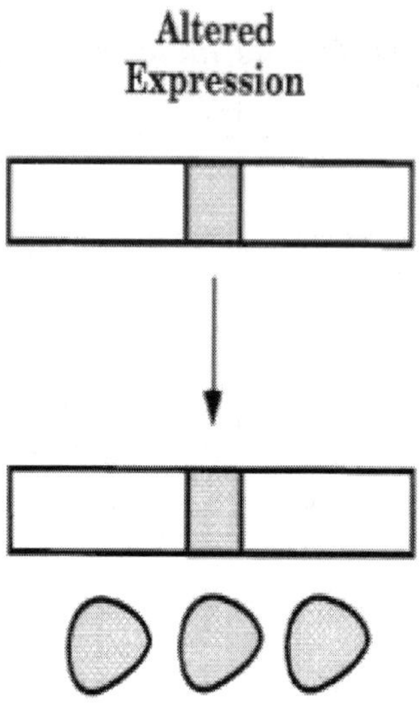

Figure 9.8. Molecular mechanisms of resistance (Altered expression).

Source: https://cals.arizona.edu/classes/ento415/LECTURES/ENTO415_Resistance.pdf

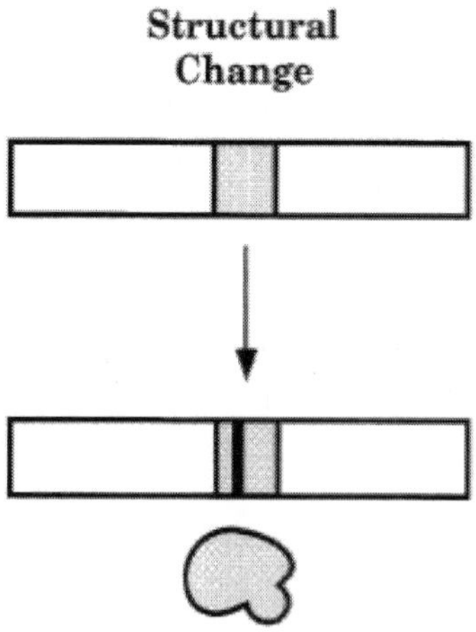

Figure 9.9. Molecular mechanisms of resistance (Structural change).

Source: https://cals.arizona.edu/clsses/ento415/LECTURES/ENO415_Resistance.pdf

Filgures 9.7 to 9.9 explains various molecular mechanisms of resistance. Figure 9.7 explains the role of estrases in conferring resistance. Figure 9.8 shows pP450, GST and esterase role in resistance. Figure 9.9 shows the role of esterases and target sites.

The insecticidal resistance can be incurred due to speeding up of the evolution process. In other words, we can say that there is an existence of microevolution. If there is much pressure due to the selection of nature, the survival will be restricted to the individuals which are considered fit and can adjust to the changing environments. This yielded some species which are resistant and their genotypes were spread all along with the barriers and boundaries. The selection pressures induced due to the insecticidal application can be taken p the resistant insects. This simple process of eveolution atleast for us has become a key concept to understand the whole process of resistance.

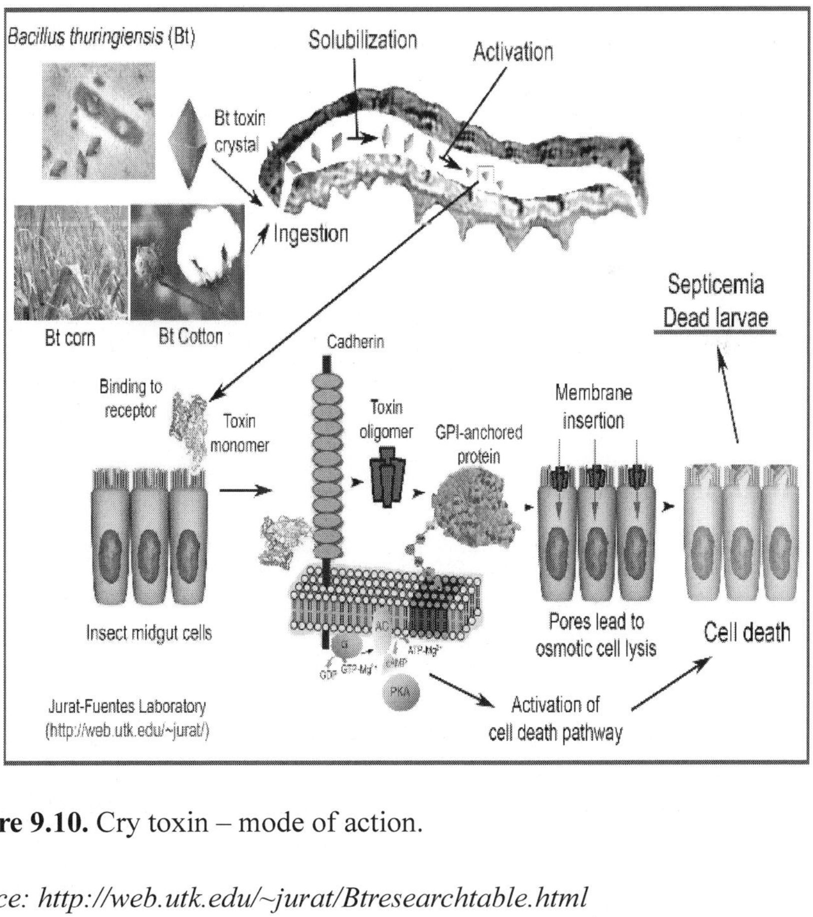

Figure 9.10. Cry toxin – mode of action.

Source: http://web.utk.edu/~jurat/Btresearchtable.html

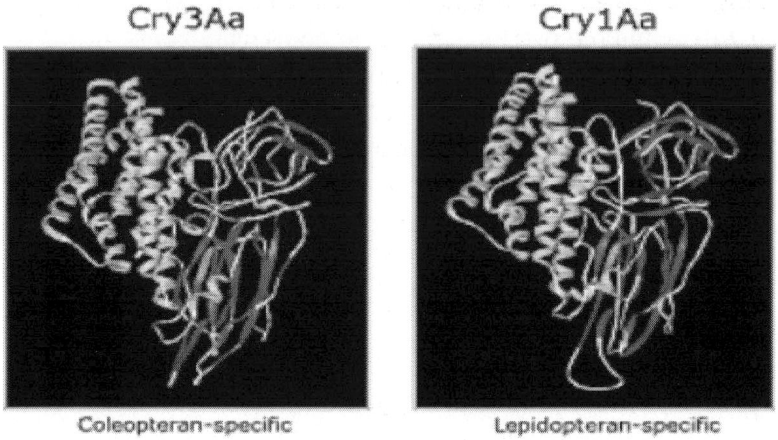

Figure 9.11. Cry toxin – Coleopteran and Lepidopteran specific.

Source: http://web.utk.edu/~jurat/Btresearchtable.html

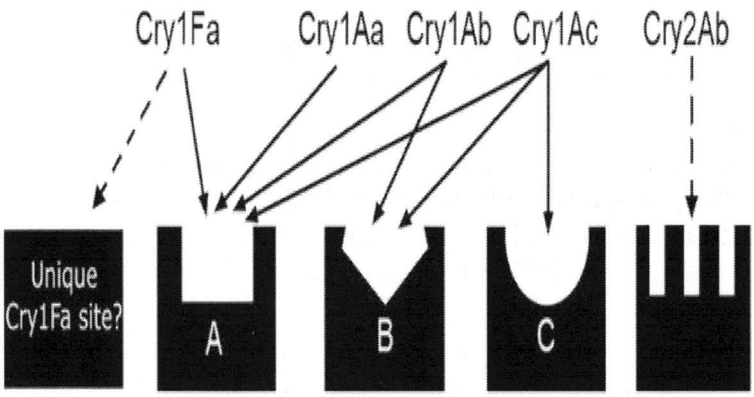

Figure 9.12. Cry toxin – Binding model.

Source: http://web.utk.edu/~jurat/Btresearchtable.html

Bacillus thuringiensiscan be named as Bt. It is classified as a Gram positive bacteria. It can produce spores. We can classify them as soil bacteria as they are predominantly found in the soils. The Bt cells can produce spores under unfavorable conditions. During the procesrulation, there will be the production of protn ceystalscrystals. These protein crystals can be calledas Cry proteins. They are basically endotoxins. They can exhibit insecticidal property. The Bt toxins were expressed in the plants through genetic engineering. The insecticidal property is a multistage process. At first, then will get ingested by an insect. The toxin will be solubulizedsolubilized. The protoxin will be altered due to the activity of the enzymes in the insects. The protoxin will get converted into a toxin. The core of the toxin can migrate to the peritrophic matrix. The receptors which are needed for this process to occur are cadherins. These receptors are dominant in the brush border membrane present in the gut. Due to this process, the oncotic pathway gets triggered or there will be the production of the toxin oligomers. These oligomers can bind with the GPI anchored proteins. They can be attracted to the lipid rafts. Dueccumulation of the toxin oligomers, there will be entryinjection of toxins inside the membranes. This will create pores in the intestine. Due to this the cells will eventually fall into osmoticock. Later the dies of internal bleeding. This process if is explained in figute figure 9.10. The cry toxins can act on both coleopteran and lepidopteran insects. These proteins were represented in figure 9.11. Figure 9.12 represents the model to explain the role of Bt toxin. The genes from one spees can be transferred into other species. This process is calledas genetic engineering.

Martineau (2001) has explained the advanced techniques used in gene transfer. Romania as rejected the proposal of planting genetically engineered plants. (James 2016). Insulin can be produced from Saccharomyces cerevisiae and E.coli (Baeshen et al. 2014). Rani and Usha in 2013 have explained the importance of modified crops. The herbicide -resistant crops and insecticidal resistant crops were produced by the process of genetic engineering (Fiester, 2006). Belgiumhas produced the first Bt crop but was not successful (Vaek et al,. 1987). Bangladsh has produced 4t- resistant Bt rinjal species (Koch et al.,2015). Bt crops which that can producedBt toxins were planted on large scale (Brookes and Barfoot, 2017). Shelton et al (2013), has explained the Bt sweet corns. Most of the crops were protected with the Bt species (Naranjo 2011). ShigetaneIshiwatarias was first to discover Bt. He named it as Bacillus sotto (Milner 1994). The bt strains were having plasma (Zakharyan, 1979).The pH of the insect gut will et increased and this reaction can kill the insect (Soberon et al,. 2010). The bt toxin can increase the pemeability of the gut for E.coli and Enterobacter (Broderiet al,. 2006). The endospoes can play a crucial role in killingthe insects (Crickmore et al,. 2014). If insects won't have eceptors, then bt toxins may not attack the gutcells (Gill et al,. 1992). Bt was produced in France(Luthy et al,. 1982). The toxin produced can separate the epithelical cells of the insects (Hofte and Whitely 1989). Btisraelensisspecies that can produce 4 Cry toxins (Ben-Dov 2014). There are about 700 Bt cry genes or proteins (Palma et al,. 2014). Most of the nucleotides are bonded together with phosphates and sugar molecules (Rettner, 017). The Bt proteins cankill coleopteran species (Chakroun et al,. 2016). Due to the presence of plasmids in bacteria, we can expect more resistance (Roh et al. 2007). Bt cotton accounts for about 19% of the total productton (James, 201). Bt cotton can produce about 6 toxins (tabashnik Tabashnicket al,. 2009). Bt toxins were reported from H.zea, Busseolafuscnd other species (Matten et al.,2008). The resistance produced intheBt cotton is due to Cry1Ab (Van Rensbur, 2007).

Bt corn species were grown in South Africa (Kruger et al,. 2009). Bt can protect from armyworms (Stewart, 2007). Cry1Ac was produced in several species (Luttrel andAli, 2007). Cry1Ac was expressed in tton in China (Liu et al,. 201. The 3rd generation of this Bthave has produced three genes namedasBollgard 2, Twin Link Plus and Wide Strike (Vyavhare, 2017). Sel crops produced Cry1Ac and Cry1Ab (Bagla, 2010). These specieshas have no harmful effect on humans and mammals (Mendelshon et al,. 2003). Some important predators are Oriusinsidiosusand Hyppodamiaconvergens(Al-Deeb and Wilde,2003). There are 22 species which can produce toxins (Naranjo, 2005). Bt cotton was used to control Bolrd (Moar et al, 2004). The pllen grains of the Bt crops can killthe insect larvae (Losey et al,. 1999). ome of the pollens never created damage to the larvae (Sears et al,. 2001).

Bt cotton can produce Cry1Ac and Cry2Ab but never affected most of the pests (Dahi, 2013). Annually Bt cotton can protect several crops (u 2010). This toxin can protect crops from ladybird (Schmidt et al,. 2009).Mealybugs and aphids were increased due to prolonged use of Bt cotton (Losey et al,. 1999). In 009, Germany has imposed a ban on Bt crops (Alvarez-Alfageme et al,. 2010). Thereare several tests to determthe effect of toxins (Hilbeck et al,. 2012). Duan et al in 2008 has have done experiments with Bt toxins an explained the role of Cry proteins.Bt toxins were not harmful to two-spotted ladybirds (Romeis et l,. 2012). Ther are some riskth these Bt toxins (Mesnage et al,. 2011). Moussa et al in 2018 has have explained the effect of aphids due to toxins. Lab tests howed that there were no harmful effects on honey bees (Rose t al,. 2007). These toxins can be accumulated in mice (Wang et al,. 2002). The risk of Bt foods is obvious (Hall, 2011).

Rats have devloped liver damage due to Bt toxins (Seralini et al,. 2007. Experiements wer done to prove the effects of Bt toxins on sheeps (Anilkumar et a,. 2010)ree Bt maize species were developed (deVendomois et al,. 2009). The Bt toxins may not be absorbedhe gastrointestinal tract of humans (CHowdhury et l.=, 2003). Here There was no systemic absorption

(Sieradzki et al,.2013). The animals were not influenced by the toxins (Herman et al,. 2006). The Cry proteins if absorbed by humans can undergo the same process as in the insects (Hammond and Jes, 2011). Seeral tests were done with rats along with famotidine (Onsone et al,. 2008. There no recombinant DNA in the gut cells (Netherwood et al,. 2004). The Bt sore prepions constitute several crystalproteins (Mezzomo et al,. 2013). lants are the same as wild plants (Domingo, 2016). Monsanto company Company has faced several problems in marketing the Bt seeds (Gucciardi, 2011). The precautionar principle concerning the Bt crops was made compulsory (Naud et al,. 2007).

REFERENCES

1. Al-Deeb MA, Wilde GE (2003) Effect of Bt corn expressing Cry3Bb1 toxin for corn rootworm on above ground non-target arthropods. Environ Entomol 32(5): 1164–1170
2. Alvarez-Alfageme F, Bigler F, Romeis J (2010) Laboratory toxicity studies demonstrating no adverse effects of Cry1Ab, Cry3Bb to larvae of Adaliabipunctata. Transgenic Res 20:467–479
3. Anilkumar B, Gopala A, Kalukumar B, Usha M, Anjaneyulu Y, Gopi KS (2010) Serobiochemical studies in sheep fed on Bt cotton plants. Toxicol Int 17(2):99–101
4. Baeshen NA, Sheikh A, Bora RS, Ahmed MM, Ramadan HA, Radwan EM (2014) Cell factories for insulin production. Microb Cell Factories 13:1–9
5. Bagla P (2010) Hardy cotton-munching pests are latest blow to GM crops. Science 327(5972):1439
6. Ben-Dov E (2014) Bacillus thuringiensisvar.israelensis and its dipteran specific toxins. Toxins (Basel.) 6(4):1222–1243
7. Bravo A, Gill SS, Soberon M (2007) Mode of action of Bacillusthuringiensis cry and Cyt toxins and their potential for insect control. Toxicon 49:422–435
8. Broderick NA, Raffa KF, Handelsman J (2006) Midgut bacteria required for Bt insecticidal activity. Proc. Natl. Acad.Sci., USA 103(41):196–199

9. Brookes, G. and Barfoot, P. 2017. GM crops: global socio-economic and environmental impacts (1996–2015). Dorcester: PG Economics LTD. Online: www.pgeconomics.co.uk/pdf/2017globalimpactstudy
10. Chakroun M, Banyulls N, Bel Y, Ferre J (2016) Bacterial vegetative insecticidal proteins from entomopathogenic bacteria. Microbiol Mol Biol Rev 80: 329–350
11. Chowdhury EH, Kuribara H, Hino A, Sultana B, Mikami O, Shimada N (2003) Detection of corn intrinsic and recombinant DNA fragments and Cry1Ab protein in the gastrointestinal contents of pigs fed genetically modified corn. J Animal Sci 81:2546–2551
12. Crickmore N (2017) Bacillus thuringiensistoxins classification, pp 41–53. http://link. springer.com/chapter/10.1007/978-3-319-56678_3.pdf
13. Crickmore N, Zeigler DR, Feitelson J, Schnepf E (1998) Revision of the nomecluture for Bacillus thuringiensispesticidal crystal proteins. Microbiol Mol Biol Rev 62(3):807–813
14. Crickmore, N.; Zeigler, D.R.; Schnepf, E.; Baum, J.; Bravo, A. and Dean, D.H. 2014.Bacillus thuringiensis toxins nomenclature. www. btnomencluture.infolifesci. sussex.ac.uk/Home/Neil_Crickmore/Bt (7 Nov. 2014)
15. Dahi HF (2013) Assessment of the effects of transgenic Bt cotton that contains Cry1Ac and Cry2Ab toxins on the abundance of non-target organisms community. Journal Nature and Science (JNSCI) 11(2):117–122
16. deVendomois JS, Roullier F, Dominique C, Seralini G (2009) A comparison of the effects of three GM corn varieties on mammalian health. Int J Biol Sci: 5(7):706-26
17. Domingo JL (2016) Safety assessment of GM plants: an updated review of the scientific literature. Food Chem Toxicol 95:12–18
18. Duan JJ, Marvier M, Huesing J, Dively G, Huang ZY (2008) A meta-analysis of effects of Bt toxins on honeybees. PLoS One 3(1):e1415
19. Fairchild EJ, ed. Cincinnati, OH: (1977) U.S. Department of Health, Education, and Welfare, Public Health Service, Centers for Disease Control, National Institute for Occupational Safety and Health, DHEW (NIOSH) Publication No. 77-180, Jul; :1-942
20. Gill SS, Cowels EA, Pietrantonio PV (1992) The mode of action of Bt endotoxins. Annu Rev Entomol 37:615–636
21. Gucciardi A (2011) Monsanto declared worst company of 2011. http://

naturalociety.com/merck-monsanto-worst-company-award, pp 1–5

22. Hall H (2011) Bt corn: is it worth the risk. http://www.scq.ubc.ca/bt-corn-is-itworth-the-risk, pp 1–3
23. Hammond BG, Jes JM (2011) Impact of food processing on the safety assessment for proteins introduced into biotechnology- derived soybean and corn crops. Food Chem Toxicol 49:711–721
24. Hammond BG, Kough J, Herouet GC, Jez JM (2013) Toxicological evaluation of proteins introduced into food crops. Grit Rev Toxixol 43(2):25–42
25. Herman R, Gao Y, Storer N (2006) Acid-induced unfolding kinetics in simulated gastric digestion of proteins. RegulToxicolPharmacol 46:93–96
26. Hilbeck A, Baumgartner M, Fried PM, Bigler F (1999) Effects of transgenic Bt cornfed prey on immature development of Chrysoperlacarnea. Environ Entomol 27:480–487
27. Hilbeck A, McMillan JM, Meier M, Himbel A, Tertikova M (2012) A controversy, revisited: is Adaliabipunctata adversely affected by Bt toxins? Environ Sci Eur 24:10
28. Hilbeck A, Moar W, Pusztai-Garet M, Bigler F (1998) Toxicity of Bacillus thuringiensis Cry1Ab toxin to Chrysoperlacarnea. Environ Entomol 27:1255–1263
29. Hofte H, Whitely HP (1989) Insecticidal crystal proteins of Bacillus thuringiensis. Microbiol Mol Biol Rev 53(2):242–255
30. James, C. 2016. Global status of commercialized Biotech/GM crops: ISAAA Brief No 52. (http://www.isaaa.org)
31. Koch MS, Ward JM, Levine SL, Baum GA, Vicini JL, Hammond BG (2015) The food and environmental safety of Bt crops. Front Plant Sci 6:283–336
32. Kruger M, Van Renzburg JJ, Van den Berg J (2009) Prospective on the development of the stem borer resistance to Bt maize and refuge compliance at the Vaalharts irrigation scheme in South Africa. Crop Port 28:
33. 684–689
34. Liu F, Xu Z, Zhu YC, Huang F, Wang Y, Zhou W, Shen J (2010) Evidence of field evolved resistance to Cry1Ac expressing Bt cotton in Helicoverpaarmigera in northern China. Pest Manag Sci 66:155–161

35. Liu XD, Zhai BP, Zhang XX, Zong JM (2005) Impact of transgenic cotton on Aphis gossypii. EcolEntomol 30:307–315
36. Losey JE, Rayor LS, Carter ME (1999) Transgenic pollen harms monarch larvae. Nature 399:214
37. Lu YH (2010) Mirid bug outbreaks in multiple crops correlated with wide scale adoption of Bt cotton in China. Science 328:1151–1154
38. Lu YH, Kongming W, Yuying J, Yuyuan G, Nicolas D (2012) Widespread adoption of Bt cotton and insecticide decrease promotes biocontrol services. Nature 487:362–365
39. Luthy P, Cordier J, Fischer H (1982) Bacillus thuringiensis as a bacterial insecticide: basic considerations and application. In: Kurstack E (ed) Microbial and viral pesticides. Marcel Dekker, New York, pp 35–74
40. Luttrel, R.G. and Ali, M.I. 2007. Exploring selection for Bt resistance in heliothines: results of laboratory and field studies. Pp. 1073–1086. In: D.A. Richter (ed.) Proceedings, 2007, Beltwide cotton conference, 9–12
41. January, 2007. New Orleans
42. Martineau B (2001) First fruit: the creation of the FlavSavr tomato and the birth of biotech foods. McGraw-Hill, p 269
43. Matten SR, Head GP, Quemada HD (2008) How governmental regulation can help the integration of Bt crops within IPM programs. In: Romois J, Shelton AM, Kennedy GG (eds) Integration of insect resistant genetically modified crops within IPM programs. Springer, New York, pp 27–39
44. Mendelshon M, Kough J, Vaituzis Z, Mathews K (2003) Are Bt crops safe? Nat Biotechnol 21(9):1003–1009
45. Mesnage R, Clair E, Gress S, Then C, Seralini GE (2011) Cytotoxicity of human cells of Cry1ab and Cry1Ac Bt insecticidal toxins alone or with glyphosphatebased herbicide. J Appl Toxicol:1–5
46. Mezzomo BP, Miranda AL, Freire IDS, Barbosa LCP, Lacava ZGM (2013) Hematoxicity of Bacillus thuringiensis as spore-crystal strains, Cry1Aa, Cry1Ab, Cry1Ac or Cry2Aa in Swiss albino mice. J. Hematol. Thromb. Dis 1:104
47. Milner RJ (1994) History of Bacillus thuringiensis. Agric Ecosyst Environ 49:9–13
48. Moar WJ, Eubanks M, Freeman B, Head G (2004) Effects of Bt cotton on biological control agents in the southeastern United States. 1st

international symposium on biol. Control of arthropods, Honolulu

49. Moussa S, Baiomy F, Abouzaid K, Nasr M, Moussa EA, Kamel EA (2018) Potential impact of host pest fed on Bt corn on the development of Chrysoperlacarnea (Neur.: Chrysopidae). Egypt. J. Biol. Pest Control 28(23):1–6
50. Naranjo SE (2005) Long term assessment of the effects of transgenic Bt cotton on the abundance of non-target arthropod natural enemies. Environ Entomol 34:1193–1210
51. Naranjo SE (2011) Empact of Bt transgenic cotton on integrated pest management. J Agric Food Chem 59:5842–5851
52. Naud, M.; Krivine, J.; Houdebine, L.M.; Kuntz, M.; Dattee, Y. and Fellous, M. 2007.No. to a moratorium on the cultivation of Gm maize. http://nonaumorattoire. fr/english.htmnonaumoratoire.free.fr/english.htm: pp. 1-3
53. Netherwood T, Martin-Orue SM, O-Donnel AG, Gilbert HG (2004) Assessing the survival of transgenic plant DNA in the human gastro-intestinal tract. Nat Biotechnol 22(2):204–209
54. Onose J, Imai T, Hasumura M, Ueda M, Ozeki Y, Hirose M (2008) Evaluation of subchronic toxicity of diatery administered Cry1Ab protein from Bacillus thuringiensis in male rats with chemically induced gastrointestinal
55. impairment. Food Chem Toxicol 46:2184–2189
56. Palma L, Muhuz D, Berry C, Murillo J, Caballero P (2014) Bacillus thuringiensis toxins: an overview of their biocidal activity. Toxins 6:3296–3325
57. Rani SJ, Usha R (2013) Transgenic plants: types. Benefits and Future J Pharmacy Res 6:879–883
58. Rettner R (2017) DNA: definition, structure and discovery. www.livescience.com/ 37247-dna, p 3
59. Roh JY, Choi JY, Li MS, Jin BR, Je YH (2007) Bacillus thuringiensis as a specific, safe and effective tool for insect control. J Microbiol Biotechnol 17(4):547–559
60. Romeis J, Alvarez-Alfageme F, Bigler F (2012) Putative effects of Cry1Ab to larvae of Adaliabipunctata -reply to Hilbeck et al., 2012. Environ. Sci Eur 24:18
61. Rose R, Dively GP, Pettis J (2007) Effects of Bt corn pollen on honeybees: emphasis on protocol development. Apidologie 38:1–11

62. Schmidt J, Braun Cw, Whitehouse IP, Hilbeck A (2009) Effects of activated Bt transgene products (Cry1Ab and Cry3Bb) on immature stages of the ladybird, Adaliabipunctata in laboratory. Arch Environ ContamToxicol 56:
63. 221–228
64. Sears MK, Hillmich RL, Stanley DE, Oberhouser KS, Pleasants JM (2001) Impact of Bt corn pollen on monarch butterfly populations: a risk assessment. Proceedings of the National Academy of science (PNAS) 98:11937–11943
65. Seralini GE, Cellier D, de Vendomois JS (2007) New analysis of a rat feeding study with a genetically modified maize reveals signs of haepatorenal toxicity. Arch Environ CotamToxicol 52(4):596–602
66. Shelton AM, Olmstead DL, Burkness EC, Hutchison WD, Dively G, Welty C (2013) Multi-state trials of Bt sweet corn varieties for control of the corn earworm. J Econ Entomol 106:2151–2159
67. Sieradzki Z, Mazur M, Kwiatek K, Swiatkiewicz S, Koreleski J (2013) Assessing the possibility of genetically modified DNA transfer from GM feed to broiler laying hen, pig and calf tissues. Pol J Vet Sci 16:435–441
68. Smith JM (2007) Genetic roulette: the documented health risks of genetically engineered foods, Fairfield
69. Soberon M, Pardo L, Monoz C, Sanchez J, Gomez I, Porta H (2010) Pore formation by toxins. In: Andeluh G, Lakey J (eds) proteins: membrane binding and pore formation. Landes bioscience and springer science, New York, pp 127–142
70. Stewart, S.D. 2007. Bt cotton. www.utcrops.com/cotton_insects/pubs/Wi29-Btcotton
71. Tabashnik BE (1994) Evolution of resistance to Bacillus thuringiensis. Annu Rev Entomol 39:47–49
72. Tabashnik BE, Van Renzberg JJ, Carriere Y (2009) Field-evolved insect resistance to Bt crops: definition, theory, and data. J Econ Entomol 102(6):2011–2025
73. Vaek M, Reynaerts A, Hofte A (1987) Transgenic plants protected from insects. Nature 325(6125):33–37
74. Van Rensburg JJ (2007) First report of field resistance by stem borer, Busseolafusca to Bt transgenic maize. S. African J. Plant Soil 24:147–151

75. Vyavhare, S.S. 2017. Bt cotton. Lubbock.tamu.edu./2017/Bt_cotton
76. Wang Z, Wang Y, Cui H, Xia Y, Altosar J (2002) Toxicological evaluation of transgenic rice flour from Bacillus thuringiensis. J Sci food and Agriculture 82(7):738–744
77. Zakharyan RA (1979) Plasmid DNA from Bacillus thuringiensis. Microbiologia 48(2): 226–229
78. P., D. B. K., &Geiling, E. M. (1959). Textbook of toxicology. New York: Oxford Univ. Press.
79. Stibitz, G., Brody, T., & Schroeder, Edison. (1966). LETTERS. Scientific American, 215(5), 8-17. Retrieved February 28, 2020, from www.jstor.org/stable/24931320
80. Ames, B. N., Mccann, J., & Sawyer, C. (1976). Mutagens and Carcinogens. Science, 194(4261), 132–133. doi: 10.1126/science.194.4261.132
81. Wildemauwe, C., Lontie, J.-F., Schoofs, L., &Larebeke, N. (1983). The mutagenicity in procaryotes of insecticides, acaricides, and nematicides. Residue Reviews, 129–178. doi: 10.1007/978-1-4612-5601-4_2
82. Zeidler, M. D., Gill, D. S., &Zeidler, M. D. (1992). Molecular Reorientation of Methanol in Mixtures with Carbon Tetrachloride. ZeitschriftFürNaturforschung A, 47(7-8), 857–864. doi: 10.1515/zna-1992-7-810.
83. Staudinger, H., &Ruzicka, L. (1924). Insecticidal substances II. For the constitution of chrysanthemum monocarboxylic acid and dicarboxylic acid. Helvetica Chimica Acta, 7(1), 201–211. doi: 10.1002/hlca.19240070125
84. Verschoyle, R. D., & Aldridge, W. N. (1980). Structure-activity relationships of some pyrethroids in rats. Archives of Toxicology, 45(4), 325–329. doi: 10.1007/bf00293813
85. Clark, T., & Phillips, A. (2001). Pesticides in Perspective. Assesing dermal exposure to pesticides from non-agricultural uses: A UK Health and Safety Executive (HSE) perspective. Journal of Environmental Monitoring, 3(1). doi: 10.1039/b100133g
86. Nobel Prize for Medicine: Dr. Paul Müller. (1948). Nature, 162(4123), 727–727. doi: 10.1038/162727a0
87. Leck, J. B., & Park, B. (1981). A comparative study of the effects of warfarin and brodifacoum on the relationship between vitamin K1,

metabolism and clotting factor activity in warfarin-susceptible and warfarin-resistant rats. Biochemical Pharmacology, 30(2), 123–128. doi: 10.1016/0006-2952(81)90182-9.

88. Pesticides group symposium. Environmental safety of pesticides; the role of field testing. (1983). Pesticide Science, 14(6), 634–637. doi: 10.1002/ps.2780140611
89. Townsend, M. G., &Busvine, J. R. (1969). The Mechanism Of Malathion-Resistance In The Blowfly ChrysomyaPutoria. EntomologiaExperimentalis Et Applicata, 12(3), 243–267. doi: 10.1111/j.1570-7458.1969.tb02522.x
90. Laskowski, J. (1982). Interfacial Chemistry of Mineral Processing Separations. Surface and Colloid Science, 315–357. doi: 10.1007/978-1-4613-3204-6_4
91. Riggs, R. L., &Biggar, J. W. (1965). Ultrasonic Suspension ofChlorinated Pesticides in Aqueous Media1. Soil Science Society of America Journal, 29(5), 629. doi: 10.2136/sssaj1965.03615995002900050041x
92. Matsumura, F. (1975). Introduction. Toxicology of Insecticides, 1–16. doi: 10.1007/978-1-4613-4410-0_1
93. Ruzo, L. O. (1983). Photooxidation Reactions Of Pyrethroid Insecticides. Natural Products, 201–206. doi: 10.1016/b978-0-08-029223-6.50036-7
94. Schechter, M. S., Green, N., &Laforge, F. B. (1949). The Synthesis OfCyclopentenolones Of The Type Of Cinerolone. Journal of the American Chemical Society, 71(4), 1517–1517. doi: 10.1021/ja01172a531
95. Gammon, D. W., & Pfeifer, K. F. (2001). Risk Assessment for Chronic Exposure to Pesticides: The Triazine Herbicide Cyanazine. Handbook of Pesticide Toxicology, 707–725. doi: 10.1016/b978-012426260-7.50036-7

INDEX